NOBLE LUXURY VILLA DESIGN

诠释豪宅

ID Book 图书工作室　编

華中科技大學出版社
http://www.hustp.com
中国 · 武汉

Using Design Contain The Whole Life

Chinese people have a persistent pursuit on residence which stems from a permanent and sustainable adore. Since ancient times, the study of the residence has pushed the development of the residential culture.

An ideal residence not only can satisfy daily need but also is a symbol of quality life. In some people's mind, luxury villa has much space to waste, while in fact it is not true. Nowadays, luxury villa is bearing more club function except for being a complete private place. With the innovation of technology and development of times, the spirit of the past can be restored in ancient ways, meanwhile the space can be more full of concept of "Home". Louis-Ambroise Dubut once vividly described, "Are there any place more comfortable than one's home? It brings you happiness; it lets you live in happiness every day."

Past and present, simple and sophisticated, classic and current, different voices fiercely collided with each other. It seems like the Big Bang giving birth to many new spaces. In fact, design is like life-by drinking water, one knows hot and cold. Complicated consideration often fall behind the free voice in people's mind. If one has the spirit of independence and the idea of freedom, then any place can be his house. Hundreds of design is an expression of material civilization. For someone with design ideal, he can show things from the deep memory and express the attitude towards the world.

Giving up the thinking of "The so-called luxury villa" while getting into the nature of life pursuit. I am keeping on telling my group that design comes from life, only real life leads to the origin of design.

Pang Yifei
Chief of Joint Meeting of CIDEA
President of Chongqing Pinchen Decoration Construction Design Co., Ltd.
High-Class Legalized Designer of APDC International Design Communication Center
Top-Ten Influential Designer from 2012 to 2013 of China International Architectural Decoration and Design Expo

用设计装下生活的所有

中国人对于住宅有着执着的追求，那是一种长久而持续的喜爱。从古至今对它的研究推动了住宅文化的发展历程。

理想的住宅，不仅仅用来满足人们生活的功能需求，更是品质生活的象征。许多人认为豪宅有那么多的空间可以浪费，其实不然，现如今的豪宅已经不是一个完全私密的场所，它承载着更多的会所功能。随着技术的革新，时代的发展，过往的精神也可以通过复古的方式还原，同时使空间更富有“家”的概念。Louis-Ambroise Dubut曾这样动人地描述：“还有什么地方能比满足我们需求的家更令人舒适的吗？它给我们的生活带来了快乐，使我们在快乐中度过每一天。”

过去与现在、极简与繁复，古今之声激烈地碰撞，就像是宇宙大爆炸般孕育出一个个全新的空间。其实设计如生活，如人饮水，冷暖自知，许多繁杂的考虑终不敌内心那个自由之声。若有独立之精神、自由之思想便是吾舍。成千上百的设计集合是物质文明的一种表达。对于一个拥有设计理想的人而言，通过设计能呈现记忆深处的一些东西，通过设计能表达对这个世界的看法。

放弃“所谓的豪宅”的思考，投入到本质的生活追求。我一直在不断地告诉我的团队：“设计源自生活，只有真正体会过生活才知道设计的本源。”

庞一飞

重庆市室内设计企业联合会（CIDEA）联席会长

重庆品辰装饰工程设计有限公司总裁

APDC国际设计交流中心高级认证设计师

中国国际建筑装饰及设计博览会2012-2013年度十大最具影响力设计师

CONTENTS

目录

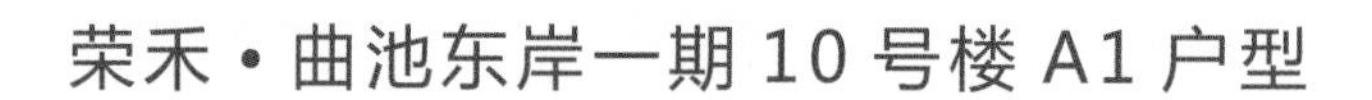

荣禾・曲池东岸一期 10 号楼 A1 户型

RONGHE•QUCHI EAST LAND FIRST PHASE BUILDING 10 # A1 UNIT

设计师：郑树芬
参与设计：杜恒、黄永京
项目地点：陕西西安
项目面积：730 m^2
文案撰写：张显梅

Designer: Simon Chong
Participatory Designers: Amy du, Jimmy
Project Location: Xi' an, Shanxi
Project Area: 730 m^2
Copywriter: Emma Zhang

THE COLLECTION OF
SIMON CHONG DESIGN

荣禾•曲池东岸A1户型，面积约为730m²，是郑树芬先生“雅奢主张”的新标杆，是最具有代表性的作品之一。荣禾•曲池东岸位居曲江核心，以高雅曼妙的姿态屹立在曲江池东岸，为西安住宅领域开拓出新的空间，重构中国历史古都雅豪圈阶层住宅区新秩序。

曲池东岸A1户型的整个空间彰显奢华和尊贵，设计师在设计方面以奢、雅、质、暖为关键词，书写了新一代雅豪们奢华生活的新篇章。

奢：雅豪之士对奢华生活之道的追求

奢，乃奢华不落俗套，体现出文化底蕴而不炫富。这是雅豪们对欣赏美、享受美的理解。项目分为三层复式，一层、二层为主要功能区，三层为视听室和棋牌室。其中一层分为客厅、中西餐厅、老人房及客房、休闲厅；二层为主卧、男孩房、女孩房、家庭厅、品茶区。四世同堂、尽显天伦之乐。客厅上层中空，挑高7.5m，尊贵大气，主卧及老人房的八角窗造型独特，完全符合雅豪之士对高品位居住空间的追求。

雅：触动高贵优雅的心

雅，乃优雅、高贵，不繁复，以及适得其所的生活理念。当今的雅豪们既具有温文儒雅的礼节，又具有高品位的文化鉴赏力和审美观，同时怀抱对历史文化的重温与铭记，怀抱对未来的畅想与人居空间的探索。因此在设计中，设计师大胆创新，以全新的手法设计出新的风格。休闲厅以高尔夫为主题，球包、书架、奖杯等饰品十分精致，茶几上饰品的摆放透露出美式的经典韵味，充分显示了一个雅豪对高品位生活的追求。

质：设计上具有较强的视觉冲击力

质，乃质感，物体的真实感，丰富多彩，做工精细，产生视觉上的冲力效果。设计师以精心挑选的材质及艺术品的搭配表现出人本设计理念。家具材质和款式均为郑树芬先生指定的BAKER品牌家具，设计上简洁大气，彰显出欧美风格。设计师选用细腻缜密的布艺、木、金属等，简洁的表象下隐藏着尊贵的内涵，陈设没有多余的造型和装饰，一切皆从功能、舒适及本真出发，空间整体气质显得更为精致尊贵。

在元素方面，以注重简美为主线，除了有软体家私之外，以经典实木家具及实木框架软装沙发做搭配。去掉了繁复的细节，简洁明快，大气有形，打造出高品质的温馨空间。

暖：春暖花开的日子

暖，乃温暖，温馨、舒适之家。这是住宅最基本的功能所需，但

11102

恰恰是设计师最难以表达和实现之处。空间层次丰富，以经典高雅的白色、冷静高贵的灰色为色彩基调，搭配温馨柔软的草绿色和明丽的柠檬黄和红色，空间尽显质朴和温馨，高贵而不俗套。客厅融入了部分中式风格，背景墙以叶片装饰挂件吸引眼球，局部柠檬黄的配色充满温暖的气息。主卧室的八角窗使视野更加开阔，暗紫色的窗帘，配合紫红色的晚礼服表达了女主人对生活的热爱。冰雪奇缘、白雪公主、小提琴、照片墙、老唱片、红酒等，无论是色彩、饰品还是主人的喜好无不体现了一种优雅浪漫的生活情调。

Ronghe • Quchi East Land A1 Unit is 730 m^2. It is a new model of "luxury and elegance" concept created by Mr. Chong. Ronghe • Quchi East Land is near the core of Qu River. It locates on the east shore of Qu River in an elegant and graceful posture. It creates a new space for Xi'an residence, reconstructing new order of Chinese ancient city's YAHAO circle.

Quchi East Land A1 Unit's whole space is luxury and noble. Designer uses luxury, elegance, quality and warmth as key words, creating new life of YAHAO's new generation.

Luxury: YAHAO's pursuit of luxury life principle

Luxury means rich and unique, showing cultural deposits without showing off wealth. This is YAHAO's understanding of beauty enjoyment. The project is three-storey construction. The first and second floors are main functional area, the third floor contains audio - visual room and chess room. The first floor room contains living room, kitchens, elder room, guest room and leisure room. The second floor room contains main bedroom, boy room, girl room, living room and tea drinking space. The four generations are under one roof, the old can enjoy family happiness. The living room's upper space is hollow, and it is 7.5 meters high full of elegant atmosphere. The main bedroom and elder room's octagonal windows have unique shape, totally satisfies YAHAO's pursuit of high-quality life.

Elegance: touching noble and elegant heart

Elegance is similar to noble, simple and suitable life concept. YAHAO has courtly etiquette, high quality cultural appreciation and aesthetic, holding the memory of history and culture, holding the thinking and search of future residence. Therefore in design, designer uses new way to design new style. Leisure room's theme is golf. Bag, shelves, trophies and other ornaments are very delicate. Ornaments display on the tea table has American classic charm, showing YAHAO's pursuit of high-quality life.

Quality: the design has high visual impact

Quality means the real sense of materials. The quality is rich and delicate, forming visual impact effect. Designer uses carefully selected materials and arts to show the human concept sense. Furniture materials and styles are BAKER's which were appointed by Mr. Chong. The design is simple and fine, showing European-American style. Designer uses meticulous technology, wood and materials. Noble connotation hides under the simple surface. The display has no odd style and decoration, all starts from function, comfort and nature. The space atmosphere is more delicate and noble.

As for the element, the space emphases on simple beauty, using classic solid wood furniture and solid wood frame sofa as collocation. Design removes complicated detail, and it is simple, high-end and warm.

Warmth: warm day

Warm is equal to comfortable, which is the basic function of residence. This is the most difficult part of expression designer. The space is rich and classic, using elegant white and noble gray as basic color, matching with soft green and bright lemon yellow and red. The space is rustic and warm, noble but not cheesy.

Living room has partial Chinese style. Background is leaf and decoration hangings. The lemon yellow color is full of warm life sense. The main bedroom's octagonal window makes the vision wider. Dark purple curtain and purple red gown express female masters' love toward life. Frozen, Snow White, violin, photo wall, old CD, wine etc. show a sense of romance and elegance from color, jewelry and masters' love.

上海银亿领墅 23C2 叠加样板房

SHANGHAI YINYI LINGSHU 23C2 OVERLAY SHOW FLAT

设计单位：上海大隐室内设计工程有限公司
设 计 师：陆凯健
项目地点：上海
项目面积：225 m²
主要材料：枫影木饰面，奥特曼大理石

Design Company: Shanghai Dayin Interior Design Engineering Co., Ltd.
Designer: Lu Kaijian
Project Location: Shanghai
Project Area: 225 m²
Major Materials: Maple shade wood finishes, Altman marble

本案是位于上海新江湾地区的联排别墅及多层叠加户型的楼盘，主要客户群为35岁~55岁的中小企业负责人。他们不仅仅满足于“居者有其屋”的最基本需求，对居住环境、住宅空间的实用性及生活品质等方面有更高的要求。

设计以简洁的西方古典元素为设计理念，以浅色效果为基调，运用特殊材质的肌理，赋予建筑精致的细部。适度运用光线来衬托空间动线，搭配现代西方古典家具，营造出具有西方古典传统风格又有现代元素的空间。

此户型为两层叠加住宅空间，通过设计使空间更加宽敞、明亮和舒适。三楼主要用于会客、日常起居。由门厅进入，在空间过渡上应用了家具及花艺，使空间通透性不受影响。餐厅墙面的壁纸由金属线条进行纵向分割，使得空间高度得以拉升，古典家具和现代的家具完美融合，蓝色花艺和墙面水蓝色油画增加了视觉效果。对应的厨房移门采用镜面，使中间过道不会因门阻隔而显沉闷。厨房内部橱柜是德国高端品牌ALNO，实木橱柜门及隐藏式的电器设备，尽显高品质。客厅墙面皮革硬包分割处理，加上大理石电视背景墙，内敛中不失奢华。共用浴室门采用明镜暗门处理，使得过道空间感更加完整。

四楼的亲子活动区内布置黑色架子鼓、海报、老式唱碟机等，在这里可以激情四溢地打着架子鼓，也可以静静听完这张老唱片。主卧室睡眠区装点浅色皮革床头软包，搭配进口丝质壁纸，开敞式更衣室及五件套豪华浴室，营造出一个高端的休憩空间。男孩

房色调为白色搭配蓝灰色，配以橘色色块家具，色彩的跳跃显露着主人的个性。客房布置温馨而又整洁，使人有宾至如归的感觉。

在材料的运用上，公共空间以相对硬性与亮面的白风影木饰面为主，私密空间则选用了较浅色系，显得明亮而宽敞。

The case is a unit of townhouses at New River Bay district in Shanghai. The major clients are middle and small company manager from 35 to 55. They are not only satisfied with the basic residence need, but also have high quality of living environment, living space's practice and life quality.

Design uses simple western classic elements as design concept, and uses light effect as basement, using special materials texture giving the build delicate details. The design properly applies light to set off the space generatrix. Using modern western classic furniture creating modern elements space with western classic traditional style.

The house is two-storey overlay residence space. Through design the space is wider, brighter and smoother. The third floor is mainly for meeting and daily life. Entering into the space, it uses furniture and flower arts, leaving the permeability unaffected.

Dining space wallpaper uses golden line to ensure straightness. The space's height has been elevated. The classic furniture and modern furniture are perfectly combined. The blue flower art and wall water blue adds the visual effect to the space. The kitchen door uses mirror, making the middle aisle seems dull for the separation of door. The kitchen inner cabinets are high-end Germany brand ALNO. The solid wood cabinet doors and hidden electronic equipment are also high qualified products. The living room wall leather hard package and marble TV background wall are introverted and not luxury. The bathroom door applies mirror door technology, making the aisle space sense more complete.

Fourth floor's parent-child activities area is set with black drums, posters, old CDs, etc., where you can passionately play the drums, or quietly listen to the old CDs. The master bedroom's sleeping area is decorated with light color leather soft bag, with imported silk wallpaper, the open plan dressing room and luxurious bathroom, creating a high open space. Boy room is white and blue-gray. Boy room and orange color furniture reveal the owner's personality. The rooms are warm and clean, people may feel at home.

On the use of materials, the public space uses relevant tough and bright White Fengying Wood as overcoating. Private space uses lighter color, which make the space seems brighter and wider.

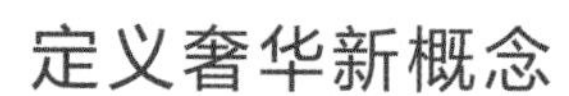

定义奢华新概念

DEFINING LUXURY NEW CONCEPT

设计单位：重庆品辰装饰工程设计有限公司
硬装设计：庞一飞、李健
软装设计：夏婷婷、黄琳
项目地点：云南昆明
项目面积：220 m²
主要材料：珍珠鱼皮、钢琴烤漆、帝壹世家壁纸、贝母马赛克、琉璃灯、钛金不锈钢、赢途石材

Design Company: Chongqing Pinchen Decoration Engineering Design Co., Ltd.
Hard Designers: Pang Yifei, Li Jian
Soft Designers: Xia Tingting, Huang Lin
Project Location: Kunming, Yunnan
Project Area: 220 m²
Major Materials: Pearl Fish Skin, Piano Stoving Varnish, Diyi Shijia Wallpaper, Fritillaria Mosaic, Color-glazed Light, Titanium Stainless Steel, Yingtu Stone

在云南，实力集团的建筑物总是流淌着一种当今难以复制的内敛和静谧。品辰设计公司拒绝无趣的装修，最大程度地做到守住这静谧的时光，打造能适应现代生活的空间。

在前期设计调研中，品辰设计小组前往魔都考察，使之在设计细节的把控上更加仔细，营造出迷人的氛围。无论是有机切割的拼花地板还是灯光设计，既彰显几何之美，又蕴含着五行八卦的韵味。采光井的增建，在室内可以自由地呼吸大自然的新鲜空气。这是一种顺畅的自然过渡，四季风光尽收眼底。

餐厅朦胧的琉璃灯在轻轻歌唱着浪漫，珍珠鱼皮演绎着尊贵气质。品辰设计公司将传统设计风格和现代元素有机结合起来，营造出一个时尚、人性化的居住空间。

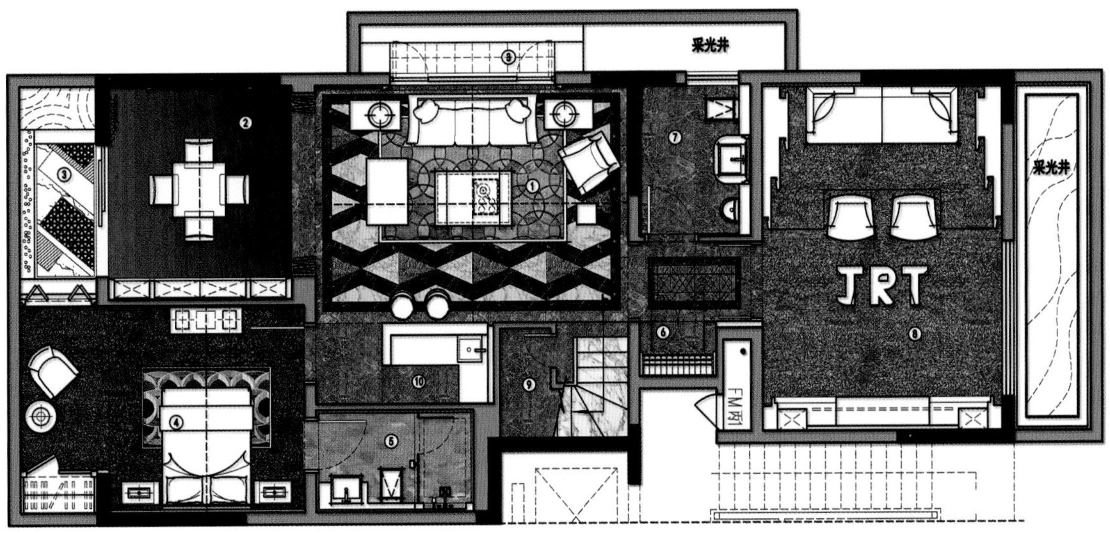
采光井
采光井
JRT
FM丙1

In Yunnan, the buildings of powerful groups always show some tranquility and restrained quality that is hard to replicate. Pinchen Design rejects decoration of no interests, trying the best to maintain this serene time and create a space fit for modern life.

In former design investigation and survey, Pinchen Design group went to Shanghai for survey, for being more specific in control of detail design and creating charming atmosphere. No matter parquet floor board of organic cutting or lighting design, all acquire geometric beauty and theory of Five Elements and Eight Diagrams.

The building of lighting well makes the air freely goes into the interior space, with smooth natural transitions and controllable four season views.

The dim colored glaze lights in the dining hall are gently singing romantic stories and the pearl fish-skin interprets the nobility. Pinchen Design combines traditional design style and modern elements, the project displays Pinchen Design's fashionable and individual living space.

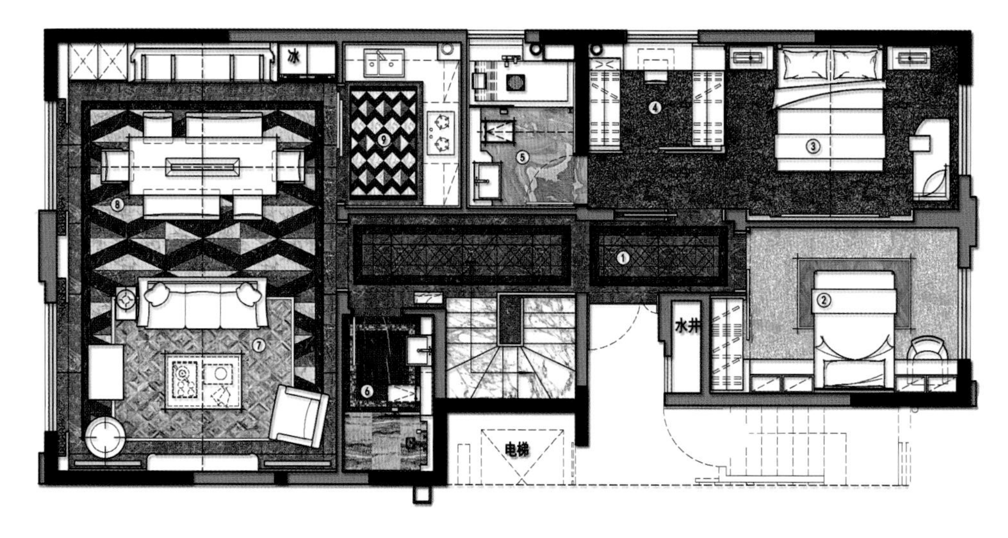

霖峰壹号样板房

LINFENG 1# SHOW FLAT

客户名称：霖峰地产
设计单位：KSL 设计事务所
设 计 师：林冠成
项目地点：广西南宁
项目面积：130 m²
主要材料：闪电米黄大理石、黑檀木纹大理石、拉丝古铜、木饰面、壁纸、银镜

Client Name: Linfeng Real Estate
Design Company: KSL Design Firm
Designer: Lin Guancheng
Project Location: Nanning, Guangxi
Project Area: 130 m²
Major Materials: Lightning beige marble, Ebony wood marble, Brushed bronze, Wood finishes, Wallpaper, Silver mirror

金、黄、红等辉煌明亮的色彩搭配，渲染出欧式宫廷不凡的气度。镶花刻金的天花墙饰与彩绘金饰的古典陈设，均以一种雍容华贵的姿态，传递着居室主人高雅的审美情趣和极富贵族文化底蕴的生活态度。该样板房的设计没有拘泥于思维的禁锢，量身定制的奢华空间，带来欧式宫廷优雅尊贵的生活感受，圆了城市新贵的宫廷贵族梦。

客厅：走进客厅，富丽的米黄大理石地面、镶花刻金的顶棚墙饰、考究的经典陈设和精美的宫廷油画瞬间将人引入古典世界。华美的水晶吊灯与精致的银镜交相辉映，浓郁的欧式宫廷风情，显露出令人凝神屏息的不凡气场。

客厅：纹饰镌刻精美的雅士白壁炉，经过设计师匠心独运的处理，既保留了古典欧式的生活情怀，又巧妙地调和了现代生活的需求。

餐厅：贵族优雅瑰丽的情愫潜入餐厅的每一个角落。设计师以美的笔触，将艺术的灵感注入日常生活中，你所享受的，绝对是一部贵族品位与现代风尚相得益彰的即兴创作作品。

次卧：当柔和高雅的贵族气质牵手现代生活的舒适温馨时，重新揣摩欧式宫廷与城市新贵的节奏，寻找属于自己的生命轨迹。

主卧：璀璨的水晶吊灯，静美的宫廷油画，奢华的床品及古典的色彩搭配，这样的家将给人带来怎样的贵族遐想……

Golden, yellow, red and other bright color form an European royal sense. The golden flower-carved ceiling decoration and colorful golden classic display uses luxury atmosphere, expressing hostess' elegant aesthetics and noble cultural life attitude. The design of show flat has not been confined by thinking, the luxury space brings European style's elegant life experience, the city new noble's royal dream comes true.

Living room: When you stepping into the living room, the luxury beige marble ground, golden flower-carved ceiling decorations, noble classic display and exquisite noble paints bring people into a classic world. The beautiful crystal ceiling light and exquisite silver mirror echo with each other. The European royal style expresses a unique shocking feel.

Bedroom: The decoration on the white fireplace is exquisite. Through the designer's creative deal. The room maintains classic European life and satisfies the demand of modern life.

Dining room: The noble's elegant emotion is full of every corner of the

dining room. Designer injects beautiful art inspiration into daily life. What you enjoy is an instant creation work of noble and modern fashion.

Additional bedroom: This is a combination of elegant noble sense and modern cozy life. The house design idea is a re-thinking of European palace and city new noble style, finding a new life pattern.

Master bedroom: Brilliant crystal ceiling light, beautiful royal painting, luxury bed products and classical color, you can imagine how great the aristocracy thinking the room brings to people!

阳台一
BALCONY
7.5m²
阳台三
BALCONY
1.6m²
1700*2000
TV
主卧
BEDROOM
13.7m²
衣帽间
CHANGE ROOM
3.8m²
厨房
KITCHEN
4.4m²
客餐厅
LIVING AREA
34.8m²
主卫
BATH ROOM
4.2m²
毛巾架
屏风推拉门
书房
STUDY ROOM
10m²
儿童房
BEDROOM
12.2m²
抽屉
抽屉
1350*1900
毛巾架
卫生间
BATH ROOM
3.9m²
TV
1500*2000
老人房
BEDROOM
13m²

巴黎密语

PARIS SWEET WORD

设计单位：大墅尚品
施工单位：大墅施工
软装设计：翁布里亚软装机构
项目地点：江苏常熟
项目面积：400 m^2
主要材料：地砖、地板、壁纸、乳胶漆、白亚光油漆、护墙板、大理石

Design Company: Da Shu Shang Pin
Construction Company: Da Shu Construction
Soft Decoration Design: Umbria Soft Decoration Institution
Project Location: Changshu, Jiangsu
Project Area: 400 m^2
Major Materials: Tiles, Flooring, Wallpaper, Paint, Bai Yaguang Paint, Wainscoting, Marble

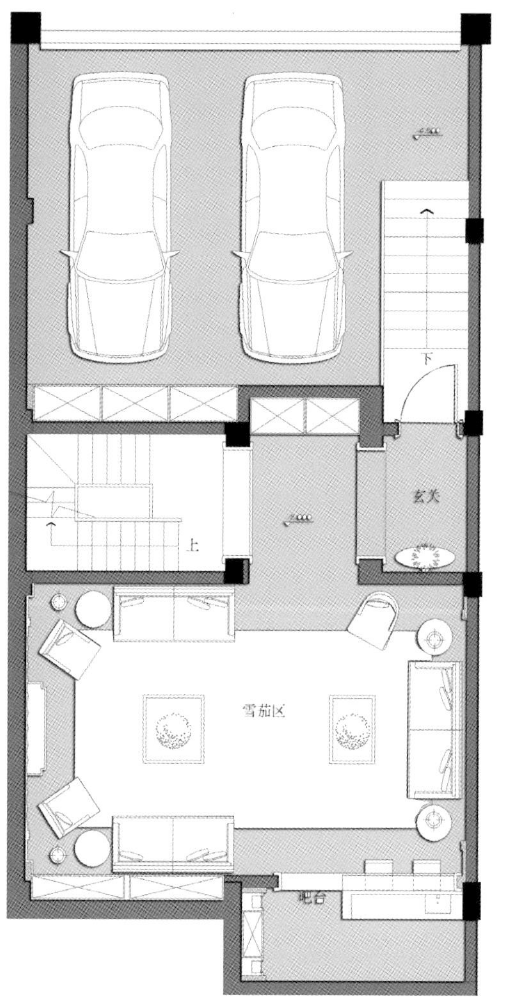
下
玄关
上
雪茄区
吧台

法国的香水、时装与美食举世闻名。这个国家给我们的感觉是充满了浪漫、精致、高贵和华丽，是一个时尚、传统与奢华并存的国度。本案业主追求居住空间的时尚、新潮和精雕细琢。我们按照业主的要求，为其营造了一种独特而华丽的室内风格。在空间布局上，突出房屋流畅的线条、恢宏的气势和豪华舒适的特点。房屋高贵典雅，温馨而又奢华。

France is famous for its perfume, fashion and food. The country gives us a sense of romance, refinement, nobility and beauty. France is a combination of fashion, tradition and luxury. The case's owner seeks for a fashionable and elegant space. We create a special and beautiful interior style for him. On the space layout, the design emphasizes on the house's smooth line, magnificent momentum and luxurious features. The house is elegant, warm and luxurious.

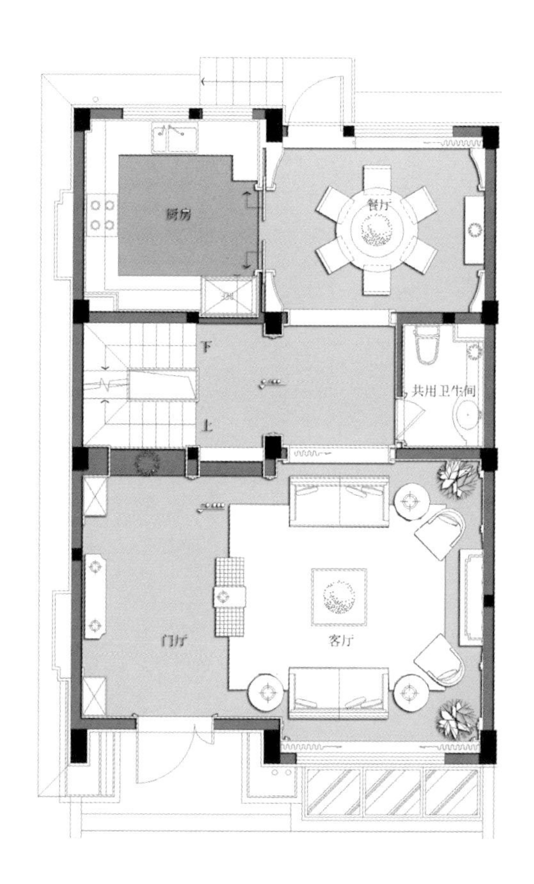

厨房
餐厅
下
上
共用卫生间
门厅
客厅

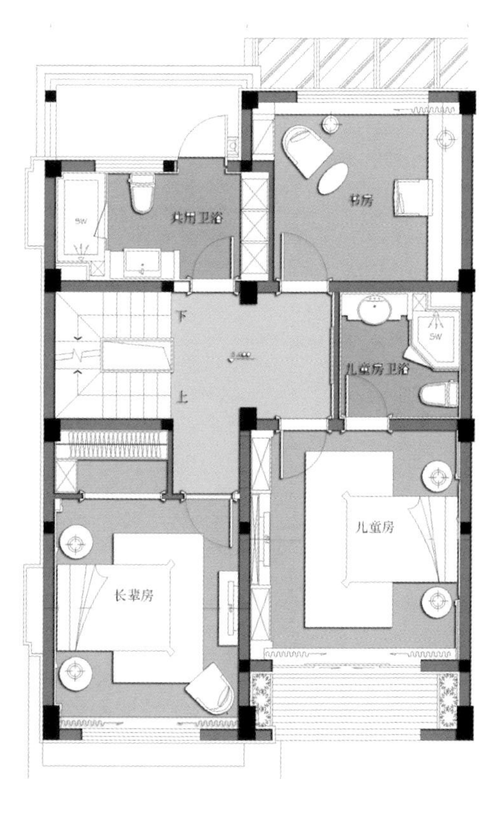

共用卫浴
书房
下
上
儿童房卫浴
儿童房
长辈房

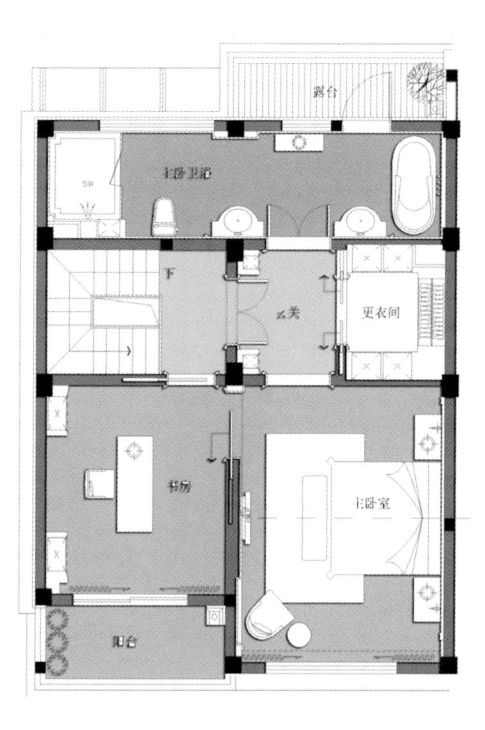

露台
主卧卫浴
下
玄关
更衣间
书房
主卧室
阳台

中航翡翠城联排别墅

ZHONGHANG EMERALD CITY TOWNHOUSES

客户名称：新疆中航投资有限公司
设计公司：PINKI 品伊创意集团 & 美国 IARI 刘卫军设计师事务所
设 计 师：刘卫军
参与设计：梁义、卢浩
陈设设计：PINKI 品伊创意集团 & 知本家陈设艺术机构
项目面积：290 m^2
主要材料：褐桃木、金钢柚地板、壁纸、大理石、艺术涂料、文化石

Customer Name: Xinjiang Zhonghang Investment Co., Ltd.
Design Company: PINKI Creative Group & American IARI Liu Weijun Designer Firm
Designer: Danfu Liu
Participatory Designers: Liang Yi, Lu Hao
Display Company: PINKI Creative Group & Zhibenjia Display Artistic Institution
Design Area: 290 m^2
Major Materials: Brown walnut, Diamond teak floors, Wallpaper, Marble, Art paint, Culture stone

欢迎参观
谢绝拍照

本项目位于乌鲁木齐市郊，为向往自然芬芳的都市精英们提供了理想的居所。室内设计从北美乡村建筑风情出发，融合当地优美的自然环境，结合乡村贵族式的生活方式，试图为居住者打造一个迷漫着自然芬芳的理想家园。

空间规划主次分明，动静分区合理，一层为客厅和用餐区，二层为睡眠和家庭交流区，三层为主人私享区，地下层为居住者提供了多元化的生活空间。

设计摒弃了烦琐和奢华，以舒适、自由为导向，强调回归自然，使空间变得轻松、舒适，散发着浓郁的自然芬芳，突出生活的闲适与自由。米黄色的质感涂料为公共空间提供了愉悦的基调，沉稳的木色带来了厚重感和舒适感，宽厚的布艺沙发温暖而惬意，点缀粗犷的石材和茂盛的绿植，散发着浓浓的田园气息。餐厅设置了田园风情的手绘油画，把用餐者带到了悠远甜美的田园之中，自然的芬芳在此得到延伸。被阳光和绿植围合起来的前院，弥漫着大自然的韵味。就连地下层也充满着阳光，一家人可以在这里喝酒品茶，观看电影，手工DIY，多元的生活方式带给居住者丰富多彩的生活情趣，营造自然的朴实，而又强调生活的舒适度，精致的细节体现着生活的优雅和从容，闲适的田园生活是每一个都市精英追求的目标。

The project is located at the countryside of Urumqi, offering ideal house to city elites. The interior design starts from North American architecture style, combining local beautiful nature environment. The countryside aristocratic life style tries to create an ideal home with natural beauty.

Prioritize spatial plan and movable and quiet districts are reasonable. The first floor contains living room and dining room, the second floor contains sleep area and family chat area, the third area is the master's private area, the basement offers a wide range of living space to occupants.

Design abandons cumbersome and luxury design, and uses comfort, freedom as orientation, and emphasizes on the nature, smooth, and strong natural fragrance, and highlights the comfort and freedom of life. Beige texture paint provides a pleasant tone to the public space. Calm wood color brings heavy and smooth sense. The sofa fabric is warm and pleasant. Rough stone and lush green plants exudes thick idyllic atmosphere. The restaurant is set up with idyllic hand-painted picture, bringing the diner to a distant and sweet countryside life. The underground floor is full of sunshine. The family can surrounded together to drink tea, watch movies, or do handmade DIY. The pluralistic way of life brings occupant colorful life fun, creating a natural and simple sense, meanwhile emphasizing life comfort. The fine details reflect life elegance and calm. The quiet rural life is every urban elites' goal.

现代中国艺术风格

MODERN CHINOISERIE

设计单位：赵牧桓室内设计研究室
设计师：赵牧桓
参与设计：施海荣、赵自强、王俊、李欣蓓
项目地点：上海
项目面积：600 m²
主要材料：水泥板、大理石、胡桃木、黑檀木、镀钛不锈钢
摄影师：李国民

Design Company: MoHen Chao Design Assoc.
Designer: MoHen Chao
Participatory Designers: Shi Hairong, Zhao Ziqiang, Wang Jun, Li Xinbei
Project Location: Shanghai
Project Area: 600 m²
Major Materials: Cement board, Marble, Walnut, Ebony, Titanium stainless steel
Photographer: Lee kuo-min

我试图用一个比较简单的形式去表达一个大都会的居住方式。我预设了两个大前提：一个是现代风格，另一个是东方的概念。对我而言，现代这个理念比较好执行，只要界定它到底是前卫，是时尚，还是相对保守就可以了。比较困难的反而是“东方”这个概念。到底“东方”意味着什么？当然这个看法见仁见智，每个人的切入点不一样，结果也就不同。但对我来说又是什么呢？其实，这一直是一个很令我困扰的问题。每当一个案子开始的时候，对我而言都是很痛苦的挣扎，因为，我必须下定决心从最初的构想中割舍很多东西，得出一个非常清楚的样板，不然的话，我永远会卡在问题里面停滞不前。那“东方”对于我究竟是什么？

最终，我决定从地面着手去解释这个问题。中国人喜欢自然的东西，这是一种文化特性。中国人喜欢搜集石头，特别是庭园景观

造景用的那些奇石。此外，中国人喜欢欣赏大理石里面自然堆砌所成就出来的如画般的天然肌理。如果把这山水般的肌理加以放大铺满整个空间，我觉得应该有点意思，索性把自己当成画匠猛往画布里泼洒墨水，地面造型就完成了。常有人问我，面对一个空间应该从哪一部分着手？平面还是立面？还是其他？我常说这没有固定答案，至少，这个案子，是从地面造型入手。解决完了地面才是平面和空间层次上的划分。完成一个案子的方法和逻辑倒是永恒不变，你还是得规规矩矩地做完平面、立面、细节、节点，等等有关的流程，它比较像是个圆形的制作流线，而不是直线般的线性工厂流水线。有时候，设计会从一个流程跳跃到另一个流程，然后回过头来再处理这个流程，这就会比较随性了。我曾经也从灯光开始发展这个空间的布局，这从来不是一个定数。

至少，我希望入口能维持住早期东方中式住宅

那种大宅门的味道，所以，我在大铁门后加上了两头镇宅的石狮子，我在石狮子后面留了开口，一方面自然光可以进入到阴暗的电梯玄关，另一方面，主人不用开门也可以看到外面的来人。

第一进的玄关是作为通往右侧公共空间和左侧私密空间的一个转折口，这是一个重要的起承转合的地方。为了让它显得比较隆重，我在其中加入了水的元素。每一个空间的链接处，我都安置了可以隐藏到墙里的条形木门，这样，主人可以自己依照特殊情况和需求分隔空间。从客厅到餐厅到收藏室都照此安排，很自然地形成该有的动线。从入口玄关往左到各个私密卧室，卧室的安排都是参照传统长幼有序的逻辑去布局的。

I try to use a relatively simple form to express a cosmopolitan way of living. I have preset two premise, one is modern style, the other one is oriental idea. For me, it is relatively easy to realize modernism, by defining whether it is advanced, fashionable or relatively conservative. However, Oriental concept is more difficult to explain. What does Orient mean? In fact, different people have different attitude towards this matter.Different pointcuts lead to different result. What does Orient mean for me? It bothers me a lot for a long time in fact. Every time when a case begin, I feel painful, because I have to give up lots of thinking in the original concept. And I have to find out a very clear image, otherwise, I will always be stuck in the issues. So, what does Orient mean to me?

In the end, I decide to solve the problem from the floor. Chinese people like natural things, which is a cultural identity. Chinese people like to

collect stones, they like to collect stones in garden landscape design, and they like to appreciate beautiful natural texture in the piles of marbles. I think it will be more interesting if amplifies the mountain and river picture in the entire space. Therefore I imagine myself as a painter, and splash inks on the paint cloth, and finish the floor design. People often ask me, which part should be the start point in a space? Plane surface or vertical surface? Or other surface? I always say that there is no fixed answer. At least, the case should start from the floor design. After the floor design I start to design the plane surface and space layers. However, the methods and logic of a case never change. One should finish the plane surface, vertical surface, detail and other process step by step. It is like a round factory line instead of a straight factory line. Sometime, design may jump from one step to another and then turns back to the former procedure randomly. I once starts from the light to develop the layout of the space. There has no fixed way.

At least, I hope the entrance can maintain the tradition of early Oriental Chinese style house. Therefore, I add two stone lions behind the iron door. And I make a open slot behind the stone lions. Thus, on one hand the natural light can permeate into the dark elevator

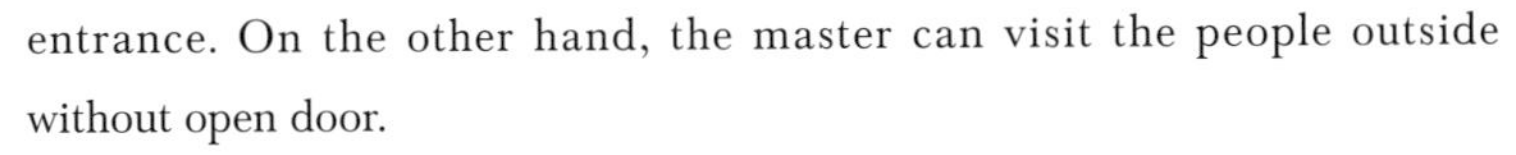

entrance. On the other hand, the master can visit the people outside without open door.

The front entrance is a turn point which links the private space on the left and the public space on the right. This is an important place. In order to make it seems more formal, I add water element into the space. At each link part of the space, I put a hidden straight wood door into the wall, therefore, the master can divide the space according to his/her own special situation. The same layout has been used in the living room, dining room, and collection room, and naturally forms generatrix. From the left of the entrance to each private room, the rooms' arrangement is in accordance with traditional senior and young order.

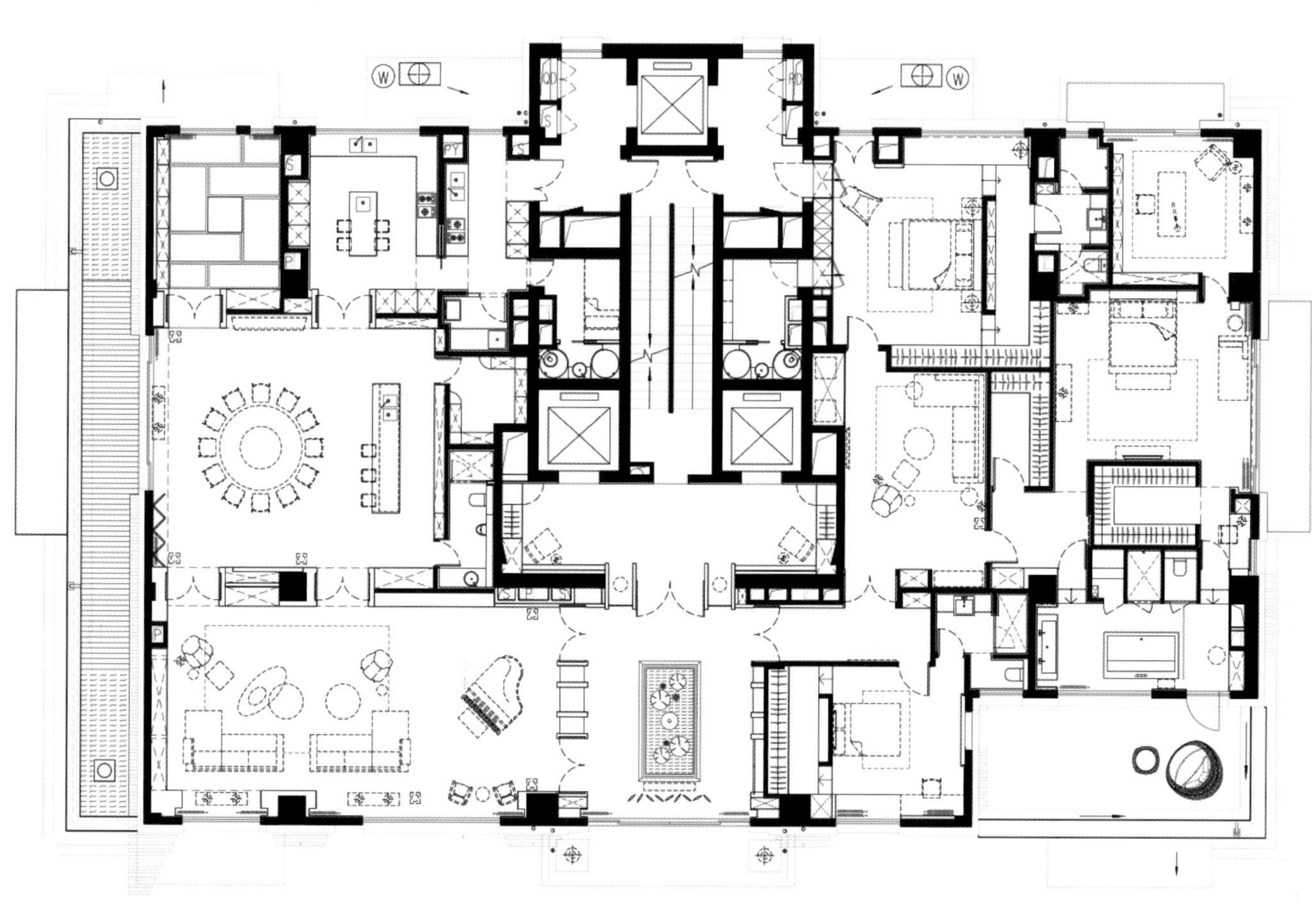

苏州盛泽湖别墅

SUZHOU SHENGZE LAKE VILLA

设计单位：大隐设计集团
设 计 师：吴正道
项目地点：江苏苏州
项目面积：490 m²
主要材料：黑胡桃木、桃花心木、大理石、壁纸、皮革、古董镜、古铜拉丝等
摄 影 师：何文凯

Design Company: Dayin Design Group
Designer: Wu Zhengdao
Project Location: Suzhou, Jiangsu
Project Area: 490 m²
Major Materials: Black Walnut, Mahogany, Marble, Wallpaper, Leather, Antique Mirrors, Antique Copper Wire, etc.
Photographer: He Wenkai

入厅玄关和客厅餐厅：玄关处挑高达到4.5m的圆形穹顶和地面精美拼花相互映衬，配以米色石材底深色木饰面，勾勒出的金属线条烘托出豪华大气的效果。客厅与餐厅相互贯通，使空间具有整体感和宽敞感，客厅主墙面运用大理石石材和深色木皮，延伸至二楼空间，并配以华丽的水晶吊灯，搭配墨绿色丝绒和蓝绿色皮革软包布艺沙发，撞色的造型窗帘，精美的镜面装饰，制作精良的雕塑工艺品，衬托出客厅的大气、奢华气氛。餐厅家具烤漆部分同客厅硬装木饰面相呼应，突出软装和硬装的整体感，同时表达出男主人对奢华生活的追求。

地下室：地下室顶部深色木饰面做九宫格分割，主墙面、镜面用深色木线条装饰，清晰流畅。棋牌室及酒吧台，运用软装装饰，宽大的组合沙发采用粗犷的皮革铆钉加雕花绒面，彰显出高贵，营造出一个温馨而又舒适的私人休闲会所。

WELCOME
ABOARD

三楼、二楼的老人房和儿童房：老人房从居住客户的心里需求出发，以舒适轻松为主，选择了相对安静的颜色——咖色和暖米色。儿童房突出表现小男孩的个人兴趣爱好，展现他热爱航海、手工制作帆船模型的特点，因此，制作了帆船船箱床等家具。

四楼、三楼主人房和起居室：主卧连接起居室各个功能区域。金色线条地毯，水晶珠帘吊灯及深紫色雕花绒窗帘等装饰着主卧的深色木饰面和皮革硬包，混搭着宽线条植绒面料壁纸，并配以游艇式黑檀钢琴漆卧床，彰显了男主人的稳重和尊贵。

Entrance into the hall and living room and dining room: The 4.5-meter height at entrance of the dome and beautiful parquet floor are echoing with each other. This material and the beige stone dark bottom wood finishes express a luxurious atmosphere with metal lines. Living room and dining room are interconnected, the space has overall sense and spaciousness. The main living room wall uses marble stone and dark wood veneer, extending to the second floor of the space, and with a gorgeous crystal chandelier. The dark green velvet and blue-green leather soft bag fabric sofa, curtain styling, exquisite mirror decoration, well-made sculpture crafts, setting off the living room's luxurious atmosphere. Paint part of the dining room furniture is echoing with the living room hard decoration wood surface, highlighting the overall sense of soft and hard equipment, expressing the host's pursuit of luxury life.

Basement: Basement top dark wood finishes are divided by Jiu Gongge. Main walls and mirrors are decorated with clear and smooth wood lines. Chess room and bar sets use soft decoration. Large sofa uses rough suede rivet and carving suede, highlights elegance, creates a warm and comfortable private lounge.

The third floor and the second floor' s elder house and children house: The elder house starts from customers' needs, choosing quiet colors — coffee and warm beige color to create peaceful and relax atmosphere. The children room shows the boy's personal interests, showing his love of sailing and hand-made sailing model. Therefore, we produce a sailing boat box-beds and other furniture.

The fourth floor and the third floor' s master bedroom and living room: the master bedroom connects all the functional area in the living room. Gold line carpets, crystal chandeliers and dark purple carved velvet curtains and etc. decorates master bedroom' dark wood finishes and leather hard pack. The wide line flocking fabric wallpaper and a yacht style ebony piano painting bed highlight the master's steady and honor.

长乐香江国际王公馆

CHANGLE XIANGJIANG INTERNATIONAL WANG GONGGUAN

设计单位：福建国广一叶建筑
装饰设计工程有限公司
设计师：唐垄烽
参与设计：叶斌
项目面积：600 m²
主要材料：意大利拿铁大理石、美克美家家具、基汀尼灯、名家窗帘、混油白色墙板、仿古镜、欣旺壁纸、美国本杰明乳胶漆等

Design Company: Fujian Guoguang Yiye Architecture Decoration Design Engineering Co., Ltd.
Designer: Tang Longfeng
Participatory Designer: Ye Bin
Project Area: 600 m²
Major Materials: Italian latte marble, Markor furniture, Ji Tingni lights, Famous curtain, Contaminated white wall, Antique mirror, Xin Wang wallpaper, American Benjamin latex, etc.

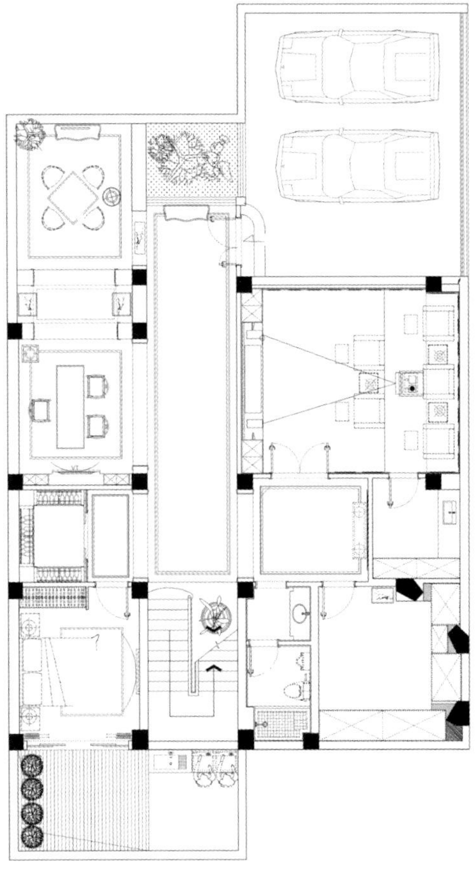

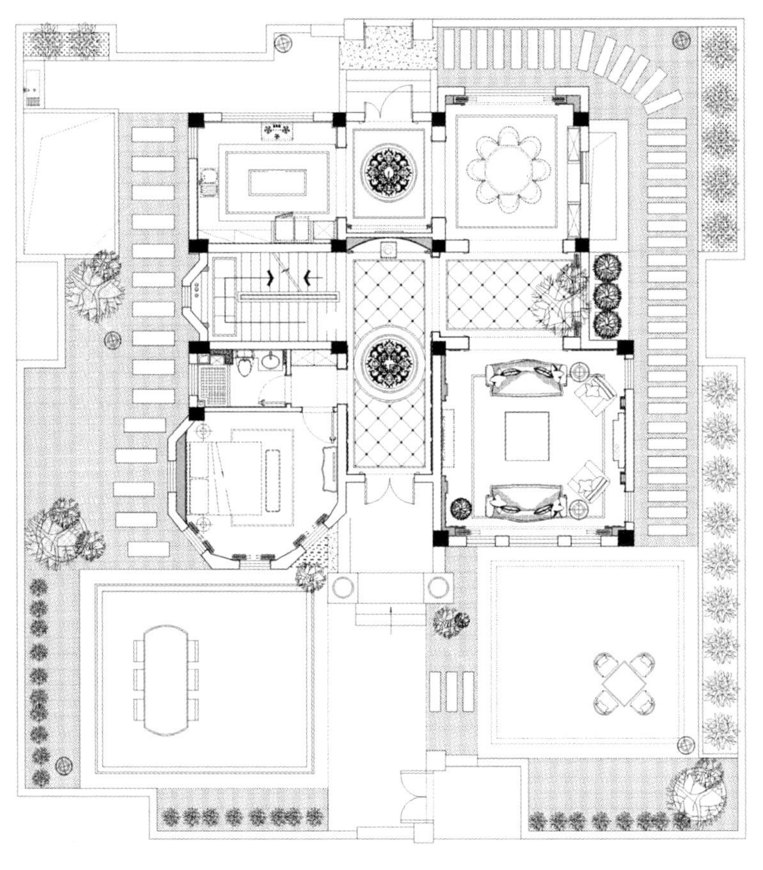

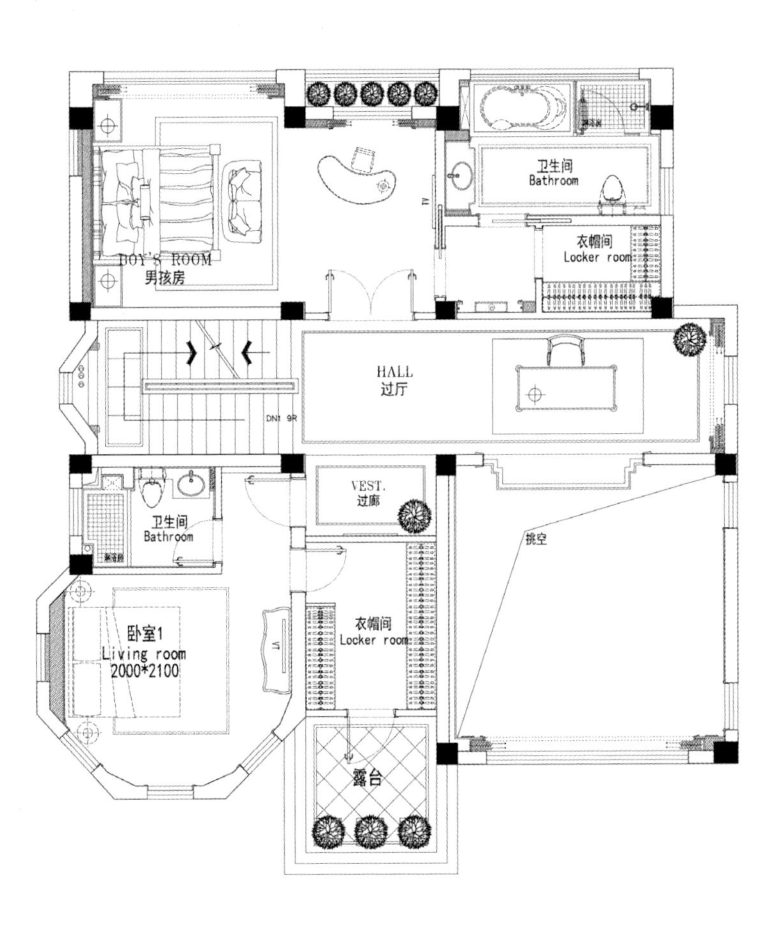
卫生间
Bathroom
衣帽间
Locker room
BOY'S ROOM
男孩房
HALL
过厅
VEST.
过廊
卫生间
Bathroom
挑空
卧室1
Living room
2000*2100
衣帽间
Locker room
露台

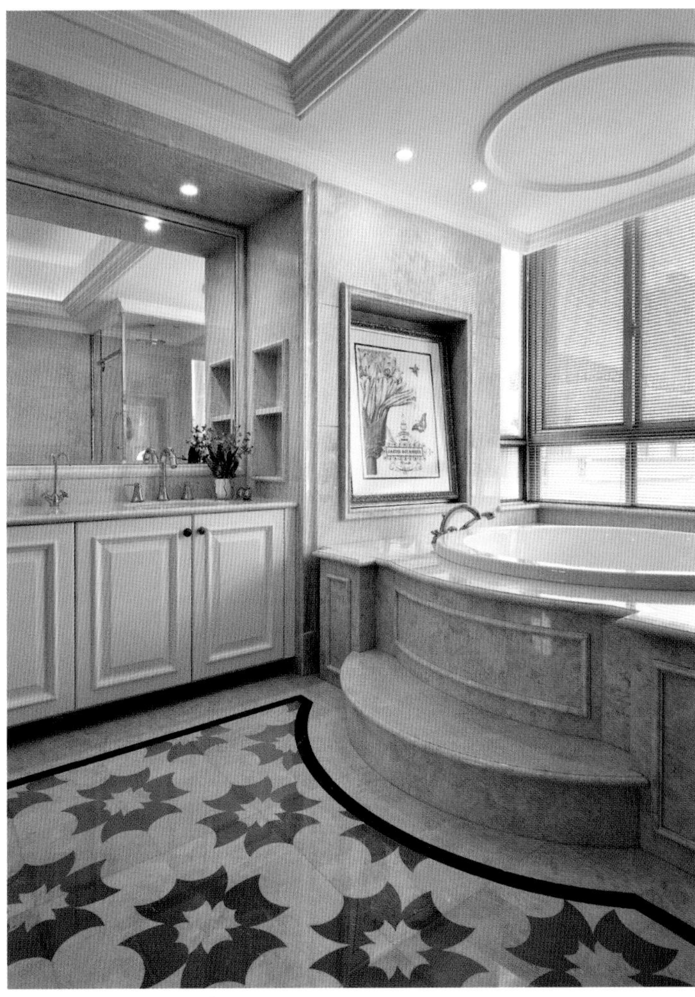

本案，美丽的别墅静静地在阳光的照耀下显得璀璨而夺目。别墅的花园中，绿草茂盛地生长，鲜艳的花朵随微风轻轻地飘摇，一切都是那么美丽，仿佛是在童话中，让人不忍触碰。

设计师选用优雅、高贵、含蓄、华丽、自然和谐为主的新古典艺术风格，搭配精致的天花吊顶、精琢的玉石梁柱、大气的挑高壁炉等，大面积运用石材提升整个空间的质感。高贵典雅的欧式造型家具使居所显得精致而富有气魄。浓烈的蓝调及皮质感更加传达出欧式风格的味道。而各区域里欧式手工沙发线条优美、颜色秀丽，注重面布的配色及对称之美，彰显居者的高贵身份，具有贵族的高贵华丽、典雅时尚的气息，让人有一种流连忘返的感觉。

In this case, the beautiful villa seems quiet, bright and eye-catching in the sunshine. In the villa garden, lush green grass grows very well, bright flowers gently sway with the breeze and everything seems so beautiful, like a fairy tale, making people can not bear to touch.

Designer chooses elegant, noble, subtle, beautiful, natural harmony neo-classical art style, with exquisite ceiling, fine-cut jade beams, atmospheric height fireplace, largely using stones to enhance the texture of the whole

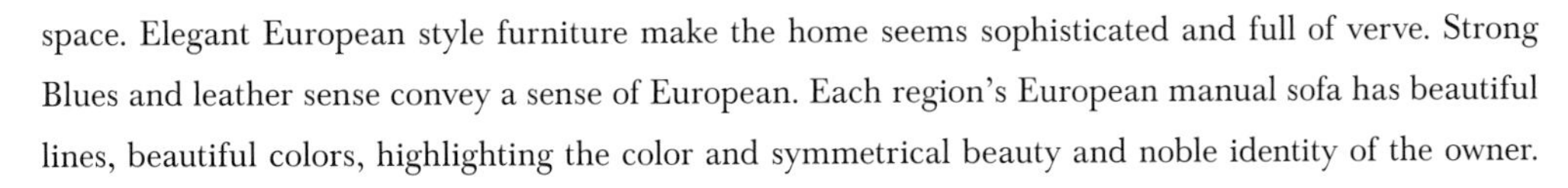

space. Elegant European style furniture make the home seems sophisticated and full of verve. Strong Blues and leather sense convey a sense of European. Each region's European manual sofa has beautiful lines, beautiful colors, highlighting the color and symmetrical beauty and noble identity of the owner.

The space has noble and beautiful aristocratic sense, making people have a feeling of nostalgia.

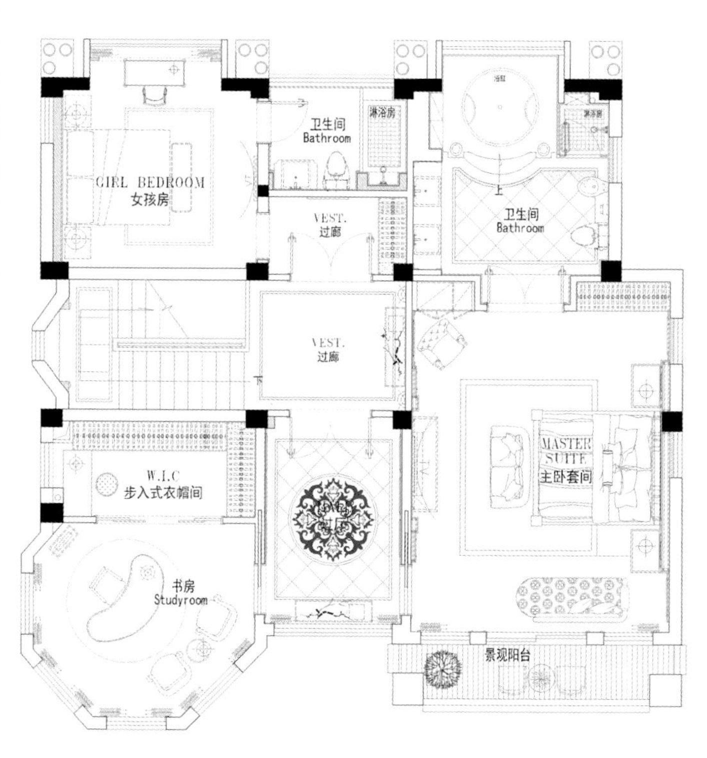
卫生间
Bathroom
淋浴房
GIRL BEDROOM
女孩房
VEST.
过廊
卫生间
Bathroom
VEST.
过廊
MASTER
SUITE
主卧套间
W.I.C
步入式衣帽间
书房
Studyroom
景观阳台

保利越秀・岭南林语 F2 别墅

BAOLI YUEXIU•LINGNAN LINYU F2 VILLA

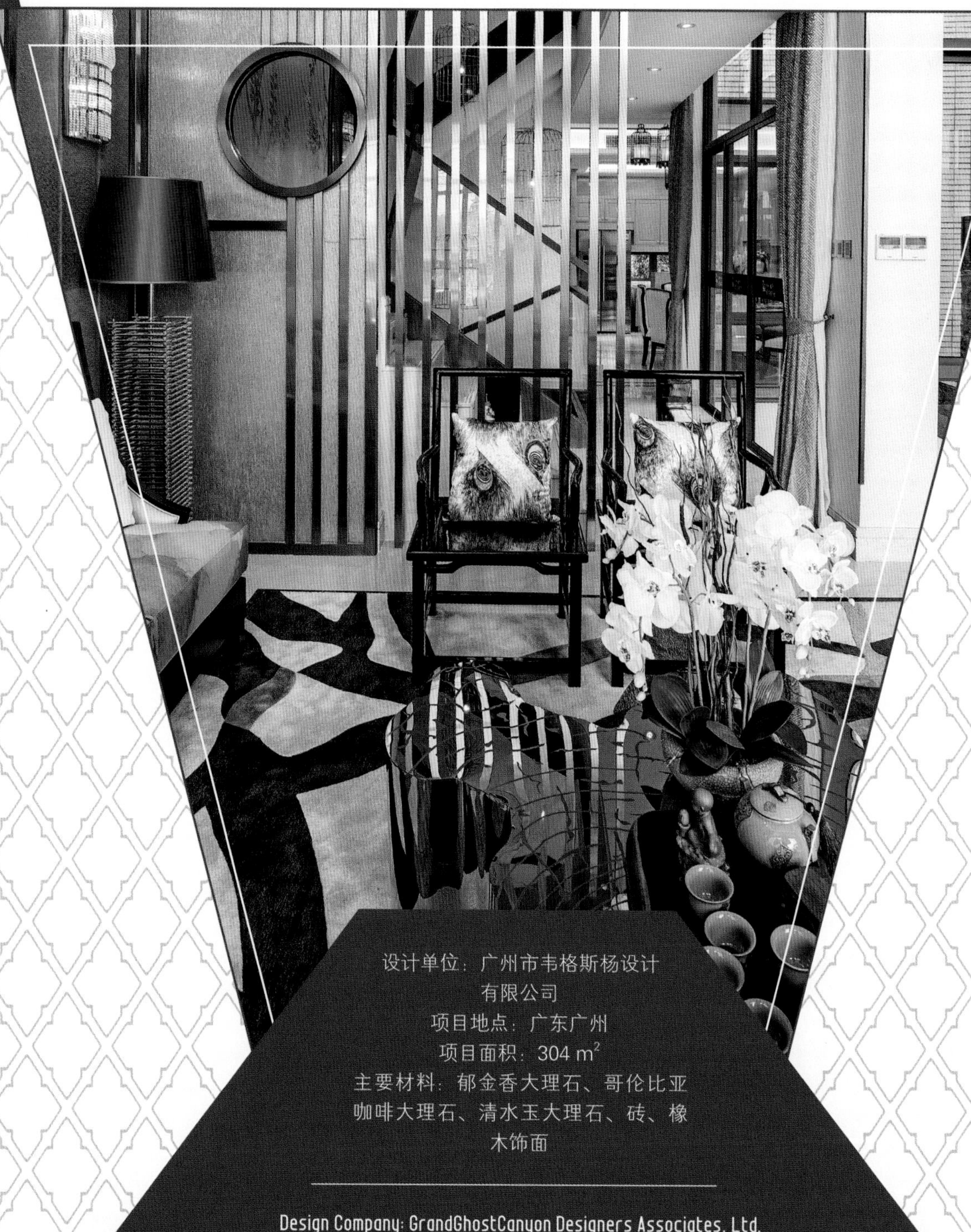

设计单位：广州市韦格斯杨设计有限公司
项目地点：广东广州
项目面积：304 m²
主要材料：郁金香大理石、哥伦比亚咖啡大理石、清水玉大理石、砖、橡木饰面

Design Company: GrandGhostCanyon Designers Associates, Ltd.
Project Location: Guangzhou,Guangdong
Project Area: 304 m²
Major Materials: Tulip marble, Colombia coffee marble, Qingshui jade marble, Brick, Oak veneer

岭南林语

本案别墅围绕内部自然山林沿山而建，零距离亲近自然山体，本户型为3层结构，设计中利用架空层和夹层的面积，使其形成一个舒适的4层别墅。

总体的设计风格采用现代中式的设计手法，将含蓄内敛与随意自然两种气质完美结合，整体基调以舒适、轻快为主，既符合人们的使用要求，又能充分展现中式的韵味。

针对居住者是古董爱好收藏者、喜欢茶乐的特点，负一层在设计上体现浓烈的文化气息，体现主人为人沉稳，喜欢中国传统文化的特点。软装上体现空间中“禅”的韵味。

首层为会客就餐区域，充分利用原有建筑形态，装饰细节上崇尚自然情趣，花鸟虫鱼，与石材等元素相结合，在现代手法的雕琢下，体现出中国的美学精神。

二层为卧室，设置有长辈房和儿童房，体现一家人其乐融融的气氛。三层为主人房，作为主人的私密空间，舒适性是关键的要件之一，通过简练的线条、软包等结合，展现一个内敛而舒适的睡眠空间；软装搭配上具有中式文化的元素，体现居住者宁静的人文空间。阳光花房，让使用者的居住空间得以从室内延伸到室外，更为女主人提供了一个独立的活动场所。

The villa built around internal natural forest along the mountain, getting close to the nature. The unit is three-storey structure, using the area of overhead floor and mezzanine to make it a smooth four-storey villa.

Overall design style is modern Chinese design, combining reservation and casual nature, the whole tone is comfortable and relax. It conforms to people's requirements, and fully demonstrates the Chinese flavor.

Since the resident is an antique collector, who likes tea and music. The design of the underground floor has strong cultural atmosphere, reflecting the owner's calm, and fall in love with the traditional Chinese culture. The soft decoration shows "Zen" in the space.

The ground floor is a parlor dining area, it makes full use of original architectural forms, the decorative details are natural taste, bird, insect and fish. They are carved by modern techniques, reflecting the spirit of Chinese aesthetics.The first floor is bedroom providing with elders and children rooms, reflecting an enjoyable atmosphere of the whole family.

The second floor is the master bedroom, as the owners' private space, comfort is one of the key elements, through the simplicity of the lines, soft pack and others, showing a restrained and comfortable sleeping space. The soft decoration has Chinese culture elements, reflecting the residents' peaceful human space. Sunshine greenhouse allows users' space extending from indoor to outdoor living spaces, providing the hostess a separate place.

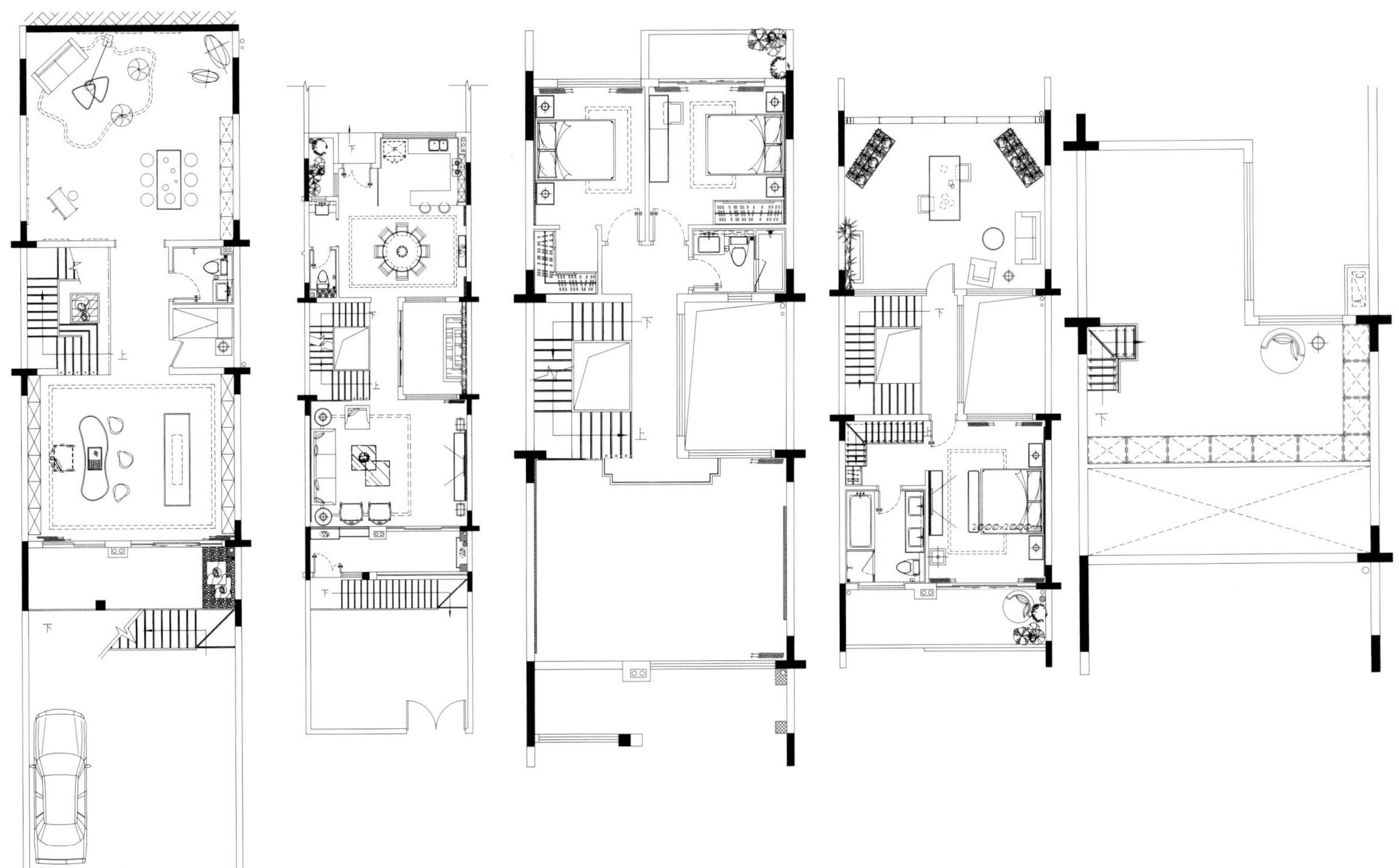

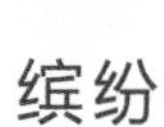

缤纷

COLORFUL

设计单位：宁波嘉德装饰工程有限公司
设 计 师：李鸣
项目地点：浙江宁波
项目面积：780 m²
主要材料：仿古砖、进口地板、进口壁纸、彩色乳胶漆
摄 影 师：刘鹰
文案撰写：王春聪

Design Company: Ningbo Jiade Decoration Engineering Co., Ltd.
Designer: Li Ming
Project Location: Ningbo, Zhejiang
Project Area: 780 m²
Major Materials: Antique Tiles, Imported Flooring, Imported Wallpaper, Colorful Latex Paint
Photographer: Liu Ying
Copywriter: Wang Chuncong

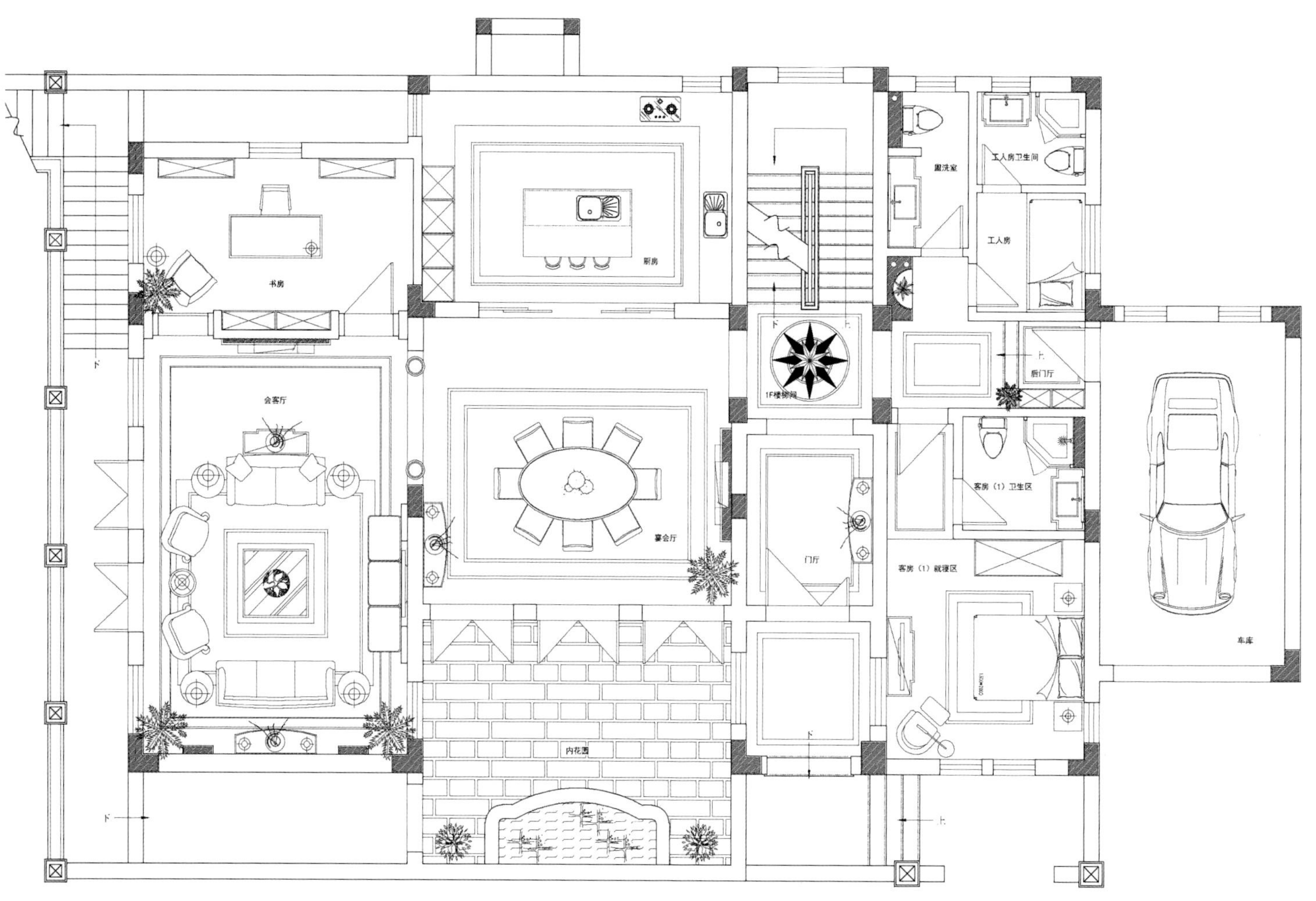

书房
厨房
盥洗室
工人房卫生间
工人房
1F楼梯间
后门厅
会客厅
宴会厅
客房（1）卫生区
门厅
客房（1）就寝区
车库
内花园

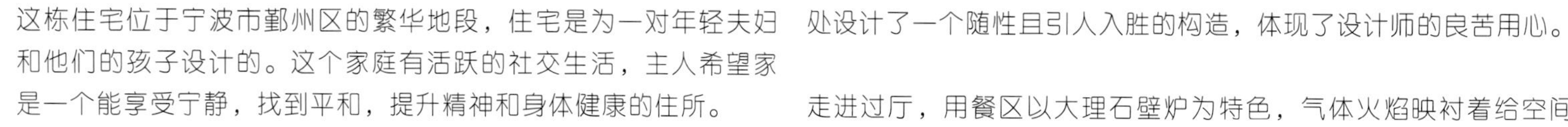

这栋住宅位于宁波市鄞州区的繁华地段，住宅是为一对年轻夫妇和他们的孩子设计的。这个家庭有活跃的社交生活，主人希望家是一个能享受宁静，找到平和，提升精神和身体健康的住所。

设计师为这栋房子设定了美式风格基调，房子内部陈设精致，玄关处设计了一个随性且引人入胜的构造，体现了设计师的良苦用心。

走进过厅，用餐区以大理石壁炉为特色，气体火焰映衬着给空间增添美好寓意的锦鲤油画，一个特色铜艺吊灯悬挂在天蓝色九宫格吊顶上。一整排玻璃门揽进了充足的自然光，使空间更宽敞

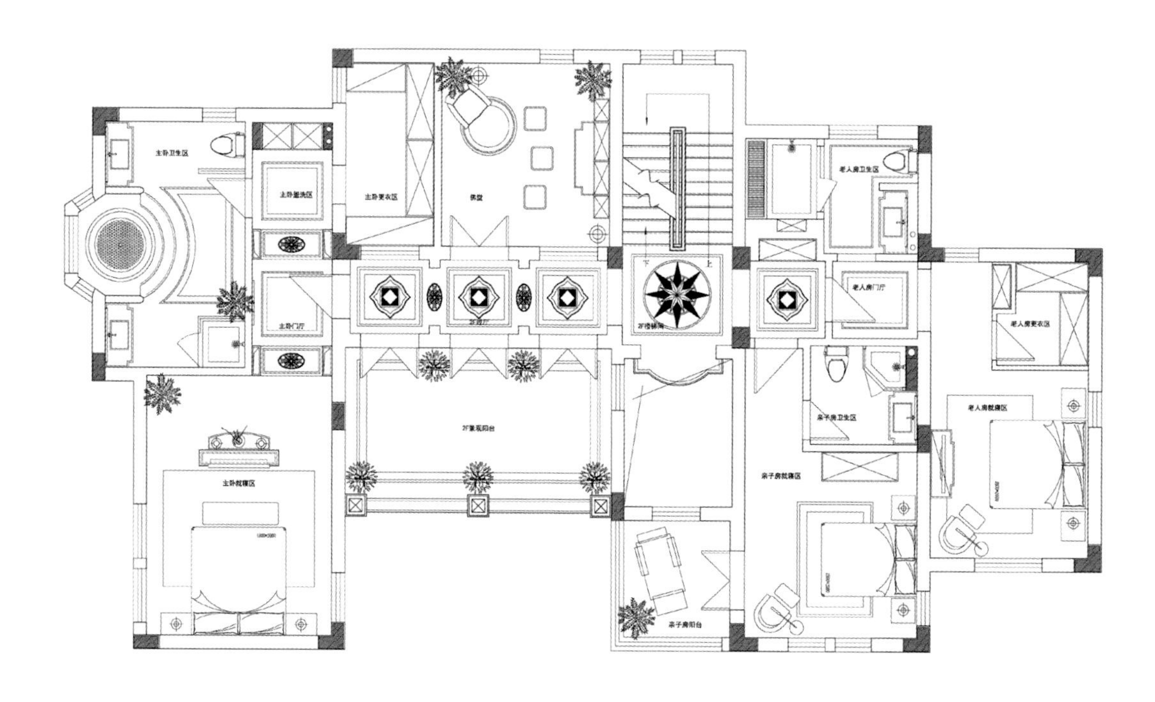

通透。客厅的白色护墙板，蓝色乳胶漆，铜艺吊灯，美式家具，拼花地砖，这些看似平常的元素搭配得相得益彰，都很好地烘托了房子的家居布置。设计师在客、餐厅间设计了一个2m多宽的垭口，在垭口两边安置了大理石罗马柱，成功放大了空间的开放感。

FINE精致家具和来自美国本土的LEXINGTON家具的完美融合，给空间添加了一种低调奢华的精致感。

地下室提供了休闲娱乐的功能，包括台球室、桌游室、影视厅、休闲茶吧等。过道的文化石背景还原了原汁原味的美式风格。各个区间不同的木饰面梁吊顶的每一次演变都给人以新鲜的体验。精美的护墙板隐藏着优质的音响系统，让人更能享受优质音乐和电影。

二楼除衣帽间、佛堂外，有三个就寝区，狭长的穹型走廊的尽头是主卧，在影影绰绰的灯光笼罩下，房屋的美感与细节十分别致，显示出主人尊贵的身份与地位。

主卧套间端庄优雅，白色护墙板，花色壁纸，绿色系窗帘，组成了一个宁静的空间，而主卧卫生间渐次递增的绿色，细腻的拼花马赛克，让人心情舒畅。而在老人就寝区与亲子区，设计师通过变换的颜色，讲究的灯光设计，增添了层次感、温

暖感和奢华感。

出于对生活的考究，设计师在二楼设计了一个能将花园景色一览无余的景观阳台。郁郁葱葱的树木，争奇斗艳的花朵，让主人在快节奏的都市生活里能够走近自然，拥抱阳光。

阁楼亲属房，吊顶采用石膏板离缝的施工工艺，与蓝白条纹壁纸完美结合，给房屋带来了积极健康的元素。

“住宅的装修细节和品质别具一格，营造了一个有魅力的生活环境。”设计师如是说。

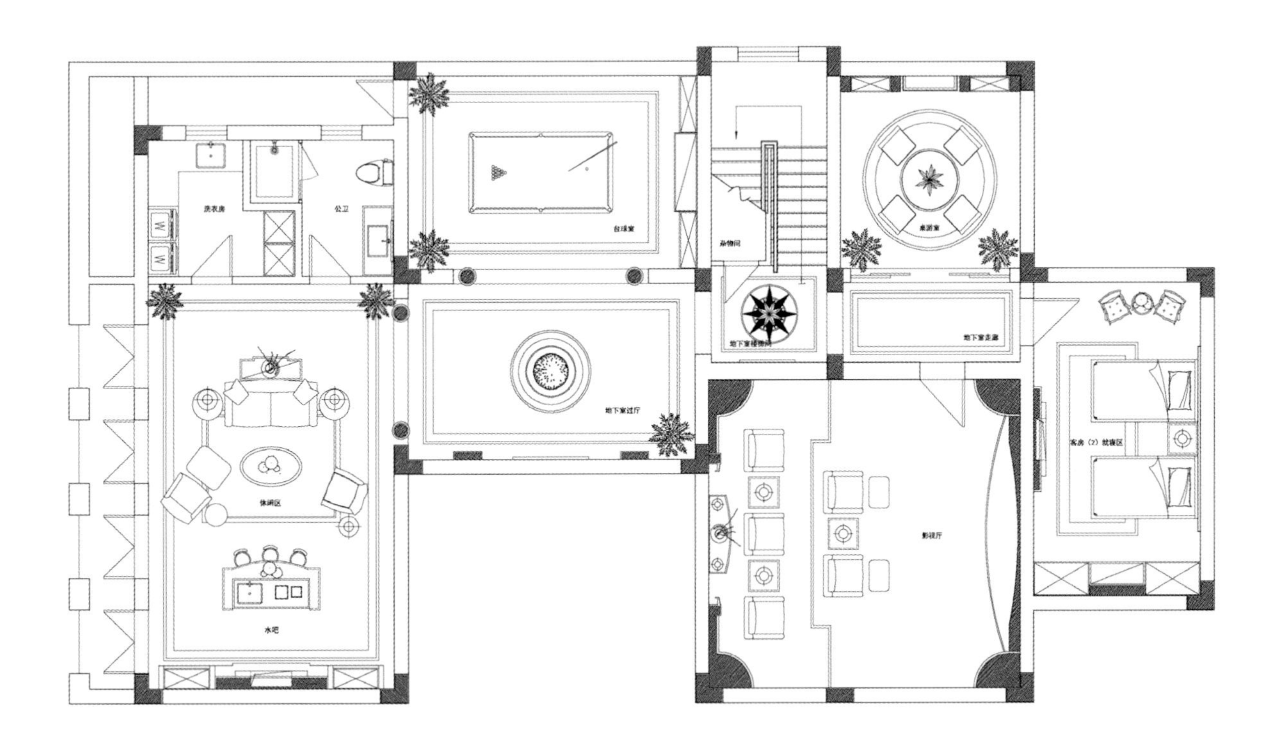

The house is located at Yinzhou, Ningbo province's famous place. The house is designed for a young couple and their child. The family has active social life. Master wishes that the house is a place to enjoy peace, find quietness and enhance mental and physical status.

Designer applies American life style tone for the house. The inner design is delicate, the porch has a natural and fascinating structure, showing designer's care and thought.
Entering into the lobby, the dining space is featured with marble fireplace. The gas flame adds good moral meaning to the space, setting off the beautiful koi painting. A featured copper light hangs on the light blue Jiu Gongge ceiling. A row of glass door collects enough natural light, making the space wider and more transparent. Living room's white panel and blue latex paint, copper light, American furniture and mosaic floor tiles are matched very well. They are well setting off the furniture. Designer designs a two meters wide bealock between living room and dining room. Both sides of the bealock are set with marble Roman column, successfully expand space's sense of openness.

FINE furniture and American local LEXINGTON furniture's perfect combination adds a kind of luxury sense to the space.

The basement offers leisure and entertainment function, including billiard room, games room, video room, casual tea area etc.. The cultural stone background restoring authentic American style. The change of different wood finishes in the various spaces leaves people fresh experience. Elegant panel hides elegant video system. Thus people can better enjoy music and movie.

The second floor has three bedrooms except for cloakroom and Buddhist temple. The end of the narrow corridors is the main bedroom. Under the light silhouettes is unique house's beauty and detail, showing master's noble identities and status.

The bedroom's suite is demure and elegant. White panel wall, flower wallpaper, green curtain combines a peaceful space. The increasing green color of the main bathroom, the detailed pattern mosaic make people feel smooth. At elder bedroom and children and parent area, designer uses change color and particular light design adds layering, warm and luxury sense.

In consideration of life quality, designer design a landscape balcony on the second floor. Lush trees and full-bloomed flower allow master getting close to nature in the fast city life.

Loft house's ceiling applies Gypsum technology and combine perfectly with blue and white strip wallpaper, brings the house positive and healthy elements.

The designer says: "House's decoration detail and quality create a charm life environment".

城市山谷别墅

CITY VALLEY VILLA

设计单位：广州共生形态工程设计有限公司
设 计 师：彭征
参与设计：彭征、陈泳夏、李永华
项目地点：广东东莞
项目面积：320 m²
主要材料：大理石、实木地板、烤漆板、硬包、不锈钢、壁纸

Design Company: Guangzhou Gongsheng Form Engineering Design Co., Ltd.
Designer: Peng Zheng
Participatory Designer: Peng Zheng, Chen Yongxia, Li Yonghua
Project Location: Dongguan,Guangdong
Project Area: 320 m²
Major Materials: Marble, Wood floors, Paint, Hard Pack, Stainless steel, Wall paper

本案针对莞深目标客户打造小户型联排别墅，项目位于广东东莞与深圳交界的清溪镇，此镇拥有得天独厚的山水资源，是一个鲜花盛开的地方。设计以“阳光下的慢生活”为主题，希望将项目的地理位置、建筑户型等优点通过样板间全方位地展现出来。

一楼的起居空间充分沐浴着明媚的阳光，室内外的空间通过生活场景的设置有效交互，尤其是室内向室外扩建的阳光房，成为传统功能的客厅与餐厅之间个性化起居生活的重要场所。

设计师摒弃客厅上空复式挑空的传统手法，使二楼的使用空间最大化。顶层的主卧不仅设有独立衣帽间、迷你水吧台，还拥有能享受日光的屋顶平台和按摩浴缸。

厌倦了都市的繁华与喧嚣后，需要一份简单与宁静。设计师摒弃了复杂的装饰、夸张的尺度及艳丽的色彩，运用了合适的尺度、明快的色调和适当的“留白”，打造出一个宽敞、温馨、舒适的空间。

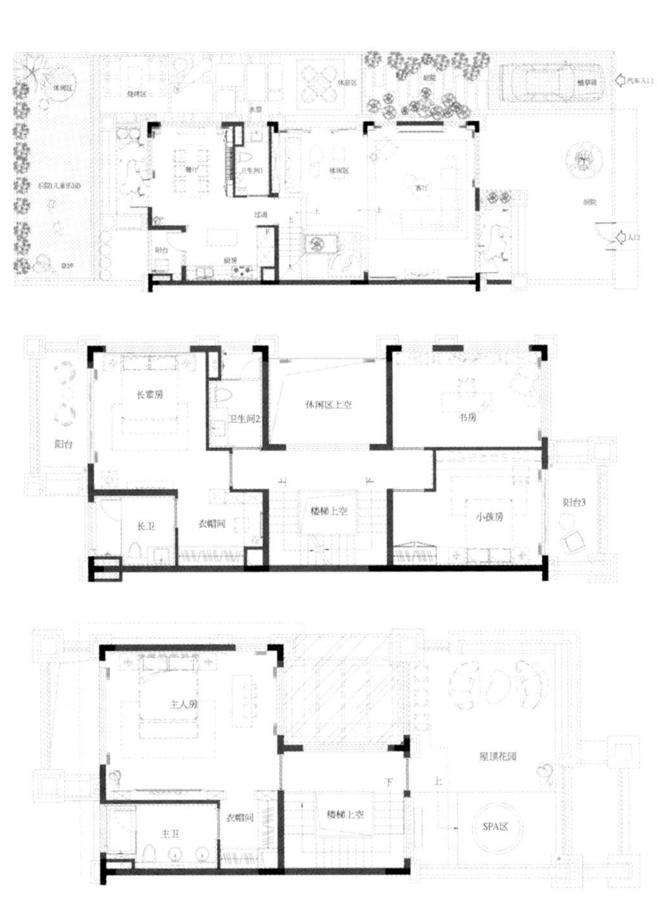

The case's target client is people in Guangzhou and shenzhen. This case is a small apartment townhouse project in Qingxi town — the convergence of Dongguan and Shenzhen. This town has a unique landscape resources, it is a place where flowers bloom. Design uses "Slow life under the Sun" as the theme, hoping to show the geographic location, construction units and other advantages through show flats.

The first floor's bedroom is in bright sunshine, indoor and outdoor spaces are effectively interacted with life scenes. The expansion of indoor and outdoor sun room becomes a traditional important place of personal living life.

Designers reject the traditional duplex to make full use of the second floor room. The top main bedroom has independent cloakroom, mini water bar and also sunny roof terrace and jacuzzi.

After getting tired with the hustle and bustle of the city, people need a piece of simple and quiet. Designer abandons the complex decoration, exaggerated scale and bright colors. They use suitable scale, bright colors and appropriate "blank", to create a spacious, warm and comfortable room.

海上国际城欧式别墅

OCEAN INTERNATIONAL CITY EUROPEAN STYLE VILLA

设计单位：上海鼎族室内装饰设计有限公司
设 计 师：吴军宏
项目地点：天津
项目面积：420 m^2
摄 影 师：上海三像摄文化传播有限公司 张静

Design Company: Shanghai Dingzu Interior Decoration Design Co., Ltd.
Designer: Wu Junhong
Project Location: Tian Jin
Project Area: 420 m^2
Photographer: Shanghai San Xiang She Culture Promotion Co., Ltd. Zhang Jing

海上国际城属于天津的空港生活配套区，本案针对的业主主要是主导社会资源运作的财智阶层。开发商有自身对建筑风格的诠释和比较严格的造价预算限制，如何在限制条件下将空间蕴含的艺术感尽情释放，并借此诠释出空间的分量，将业主的期望和开发商追求的理念相结合，都是对设计师的挑战。

设计师针对本案选择了欧式风格。欧式风格，顾名思义指的是来自于欧罗巴洲历史上的风格。业界和学术界一般认为，主要有法式风格、意大利风格、西班牙风格、英式风格、地中海风格、北欧风格等几大流派。古典欧式风格的要旨就是追求华丽、高雅的古典特性，其设计风格直接对欧洲建筑，对家具、文学、绘画甚至音乐产生了极其重大的影响，在地域之外具体可以分为以下这些：罗马风格、哥特式风格、文艺复兴风格、巴洛克风格、洛可可风格、新古典主义风格。

所谓风格，是一种长久以来随着文化的潮流形成的一种持续不断，内容统一，有强烈的独特性的文化潮流。欧式风格就是欧洲各国文化传统所传达出的强烈的文化内涵，跨越时间长河，跨越地域国别。

本案并非只限于欧式风格中的某一流派，而是有着打通各种流派的界线，形成鲜明的个性化的风格。于是我们看到了罗马式和法式的廊柱，也看到了英法风格混搭的皮椅和壁炉，更看到了欧式风格中不可或缺的水晶吊灯、镜面、新古典主义的静物油画等元素，所有这一切，将空间装扮得华彩浓烈又分外妖娆。此外，在一些精心布置的细节上也可以看到欧式古典传统源远流长的风格。但总体上而言，设计师以灵巧的匠心独具超越了欧式的烦冗和浮华，在主干部分的深色调外，尚有色调呈现出明快爽朗的一面，从而展现出了更加现代时尚的风尚气质，比如在阅读区，书架的形制就在古典中演绎出了现代大都会色彩的格调，显得摇曳多姿却又文脉悠远；在卫生间，设备本身的现代性在这个区间范围内形成了独成一体的风格。

生活在繁杂多变、喧嚣浮躁的世界里本已是烦扰不休、心绪缠绕，而温馨、自然、自由的生活空间却能让人的身心感到安静和安逸。本案借助空间结构的解构、重组、淬炼和升华，便可满足业主对悠然自得的生活的向往与追求，让业主在纷纷扰扰的现实生活中找到极致和谐的平衡，享受一个隔绝袭扰、轻松安然的温馨空间。

The ocean international city belongs to Tianjin's airport life district. The case's owners are intelligent layer leading social resource. The developers have their own explain of architectural style and strict price budget limitation. How to express the space's art sense under the limitation and show space's power and combines owner's expectation and developer's idea is the challenge of the designer.

Designer chooses European style for the case. European style comes from

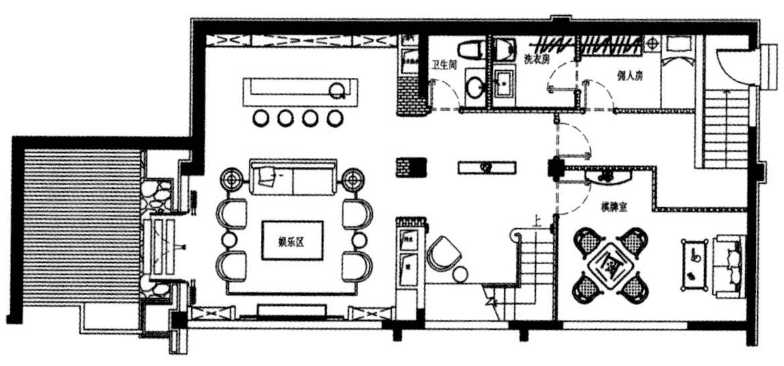
卫生间
洗衣房
佣人房
娱乐区
上
棋牌室

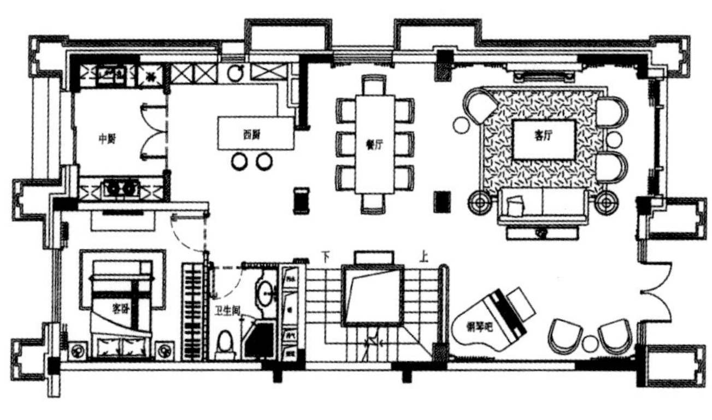
中厨
西厨
餐厅
客厅
下
上
客卧
卫生间
钢琴吧

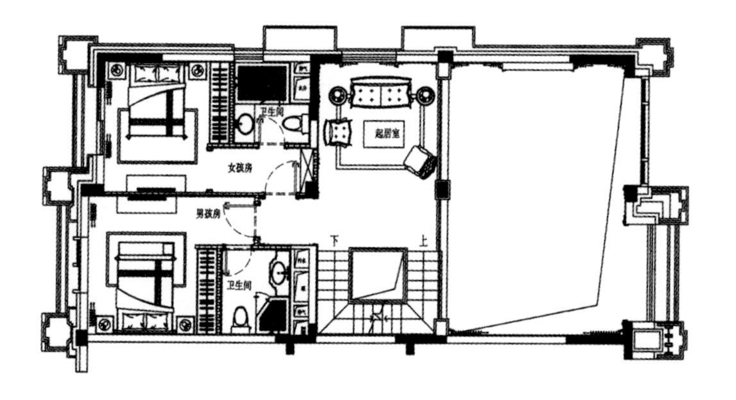
卫生间
起居室
女孩房
男孩房
下
上
卫生间

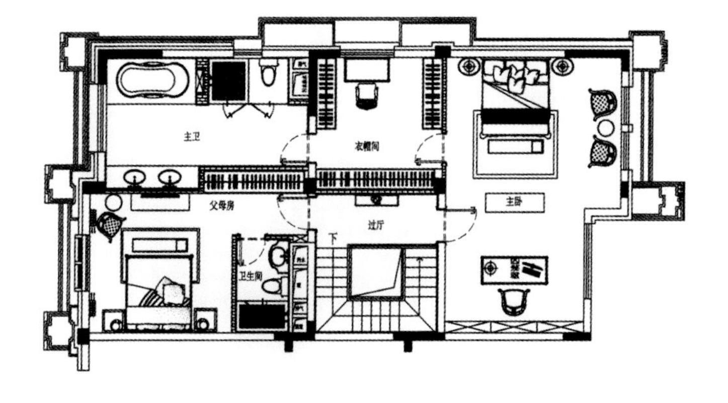
主卫
衣帽间
父母房
主卧
下
过厅
卫生间

the historical style of Europe. The study field generally thinks that there are France style, Italian style, Spain style, England style, Mediterranean style and North European style. Classic European style seeks for luxury, elegant feature. The design style directly influences on the European building, furniture, literature, paintings and music art. Outside the region, the styles can be classified as Roman style, Gothic style, Renaissance style, Baroque style, Rococo style, Neoclassical style.

Style is a continuing, unique and strong culture trends changing with the culture for a long time. The European style is a strong culture connotation of European countries' culture and tradition. This style spans time river, region and country.

The present case is not confined to a school of European style. It breaks the boundaries of various genres and forms vivid personalized style. So we see the Romanesque and French pillars, also see the British and French style leather chairs and a fireplace. One can also see the indispensable European-style crystal chandeliers, mirrors, neo-classical still life paintings and other elements. All of those decorate the space with

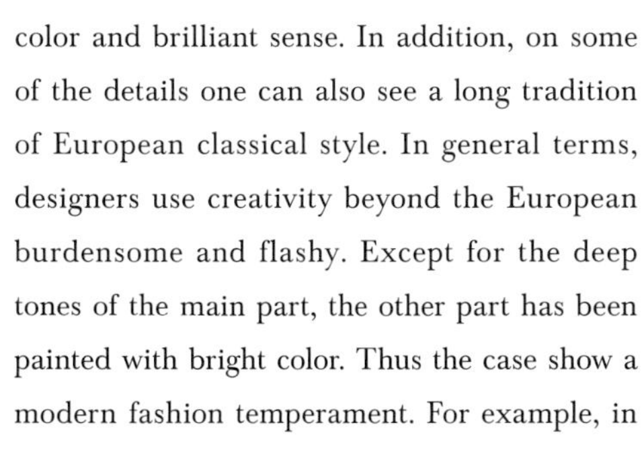

color and brilliant sense. In addition, on some of the details one can also see a long tradition of European classical style. In general terms, designers use creativity beyond the European burdensome and flashy. Except for the deep tones of the main part, the other part has been painted with bright color. Thus the case show a modern fashion temperament. For example, in the reading area, bookshelves express a modern metropolis style, seeming glittering and distant. In the bathroom, the device's modernity forms a unified style.

Living in the complex and changeable world, people always feel annoyed. Warm, natural, free living space makes people feel physically and mentally quietness and comfort. In the case of deconstruction, reorganization, refinement and sublimation, people can feel desire and pursuit of leisure life, so that the owners can find the ultimate balance in the chaos of real life, and enjoy an isolated, relaxed cozy space.

深圳半山海景别墅

SHENZHEN HALF MOUNTAIN SEA VIEW VILLA

设计单位：朗昇空间设计
项目地点：广东深圳
项目面积：1000 m²
主要材料：大理石、地砖、涂料等

Design Company: Lang Sheng Space Design
Project Location: Shenzhen, Guangdong
Project Area: 1000 m²
Major Materials: Marble, Floor Tiles, Paint, etc.

我有一所房子，
面朝大海，春暖花开
……

用知名诗人海子的《面朝大海 春暖花开》诗中这一句来形容深圳半山海景别墅设计再贴切不过了。我公司花费近一年时间设计的半山海景别墅于近期完全交付业主使用。该别墅设计共有5层，面积约为1000 m^2，建筑主体依山而建，靠山面海，森林环抱，风景秀丽。室内则空间方正、南北通透、光线充足，可谓是一所真正意义上的面朝大海、春暖花开的房子。

因别墅是旧建筑，需要重新改造，朗昇空间设计团队为保持建筑与室内的统一与协调，故将外观设计、园林设计、室内设计结合在一起来重新考虑。首先是建筑外观设计改动较大。用大理石雕刻的圆形拱门与几根巨大敦实、高高擎起屋脊的欧式石柱，显得十分稳重而有力。满铺的红褐色墙砖使建筑外形别具一格，欧式线条勾勒出建筑层次丰富的轮廓，使整个建筑外观看起来高贵、雅致，富有古典韵味。园林设计清新简洁，除了观赏鱼池及少量精心种植的小景观之外，预留了较多供家庭成员活动的空间。

客厅区域包含有主客厅、小酒吧、小会客厅等功能，由一、二楼两层贯通，楼层高、进深宽，所以客厅设计尤为宽敞大气，层次丰富。客厅背景处两根巨大的石柱，似乎是建筑元素的延伸。一楼与二楼空间之间，通过几个圆拱造形相互连接，使上下空间互动起来。顶棚造型精心雕刻，悬挂的水晶灯，光彩夺目，华丽高贵。墙面上浅黄色涂料与白色面板相间，色彩丰富温暖。精美的欧式家具、地毯、窗帘及配饰物等，均饱含着业主的个性与喜好。这些饰物大多繁花似锦，似从户外盛开到室内，然后在每个空间、每个角落又以不同的形式自然绽放。

位于客厅一侧的餐厅设计，则是用建筑内一个看似封闭的圆形区域形

成，其四周通透，光线充足，圆形的顶棚配以圆形的餐桌，顶棚上同样用雕刻的花纹构成，造型精致，其间垂吊的小型水晶灯，晶莹闪烁，搭配的黄色餐柜及灯饰，使餐厅中点缀着一丝异域的情调。

对于一个喜好欧式风格的业主来说，他似乎要求设计师不放过每一个能表现设计的机会，比如由客餐厅进入其他楼层的电梯厅、过厅、钢琴区、楼梯过道等次要空间，亦将客厅中用过的元素表现得淋漓尽致。

主卧室设计宽敞，景观视线尤佳，除摆放的双床榻之外，还摆放了一组休闲沙发茶几。主卧室设计风格与客厅区域设计几乎相同，从顶棚到床具、沙发等，亦是一番四季如春、繁花盛开的迷人景象，使主卧室充满着温暖与舒雅的气息。与主卧室配套的书房设计，在欧式风格的空间内，使用国人一贯喜欢的中式红木家具（书柜、书桌、床榻）等，散发出书香气息。

儿童房设计则显得清新淡雅，偶有几束淡淡的红色窗帘布艺，使室内充满着朝气与活力。由

于家庭生活需要，本套别墅设计有两个家庭厅，分别设计成两种风格，两个家庭厅之间通过一扇长形花边窗户连接起来。一种为与别墅整体风格保持协调的欧式风格，这种风格似适宜于家庭中年龄较长的成员，具有私密感。另一种为东南亚风格，这种风格与其他休闲空间（如棋牌室设计、酒窖设计）等联系在一起，具有浓厚的热带度假气息，色质古朴厚重，有种令人回归自然的轻松休闲感，满足了家庭成员们娱乐及对外交际的需要。

I have a house which is facing the sea
when spring is coming and temperature become warm
flowers will be blooming
......
The Half Mountain Sea View Villa Design project can be described by famous poet Haizi's poem *Facing the sea with spring blossoms*. Our company uses almost one year to design the project, now it is given to the owner. The villa has five floors, it locates 1000 m^2. The main body of the building is near the mountain, the front is facing the sea. Lush forest and landscape are beautiful. Inner space is square shape will full air and light. It is really a house facing the sea in the blooming spring.

The house is old building which need to be re-transformed. Lang Sheng Space Design makes a combination of facade design, landscape design, inner

design together to maintain the coordination of architecture and inner space. First of all, the facade has been changed a lot. Marble round door and some big and high European style column seems stable and strong. The red color brick makes the surface of the building unique. European lines stretch out rich outline of the building. The whole architecture is noble and elegant, full of classic sense. Garden design is fresh and simple, except for fish pool and some plants, it leaves more space for family members.

Living room space contains main guest room, bar, meeting room and other functions. The space has two floors. The storey is high, depth is wide. The living room design is wide and rich. The living room has two huge columns, it is the extension of architectural elements. Between the first floor and the second floor, there are some arch shape lines connecting the upper space and low space. The ceiling pattern has been carved by heart. The hanging crystal lamp is bright and noble. The wall has warm light yellow and white panel. European furniture carpet, curtain and ornaments contain owners' individuality and interests. Most of the decoration are like beautiful flowers from outer to inner, blooming in different space and different corner.

The dining room design on the side of living room is built by a closed round area in the building. The surrounding is transparent. The light is sufficient. The round ceiling matches round table. The ceiling is full of carved pattern. The room is delicate, the small chandelier on the ceiling is bright. The yellow cabinet and light add some exotic tone to the dining room.

The main bedroom is wide and has good view sight. Beside the double bed, there is a group of casual sofa table. The design style of the main bedroom and the living room are almost the same. From ceiling to bed and sofa, the flower pattern is very attracting, making the main bedroom full of warm and smooth sense. The study design uses European style. The Chinese rosewood furniture (bookcases, desks, couches, Chinese paintings) gives out book taste.

Children room design is fresh and elegant. The light red curtain fabrics makes the room full of vigor and vitality. Because of the need in family life, the captioned villa has been designed with two living rooms. They are designed in two styles respectively. The two families are connected with a long shape flower frame window. One style is European style which coordinates with the whole style of the villa, and this style is suitable for families' elder member. The style is full of private sense. Another style is Eastern and Southern style. This style is connected with other leisure space (Chess room design, Wine cellar design). The style is full of rich tropical resort ambience. The color is plain and thick, with a sense of getting close to nature, satisfying the need of family entertainment and external communications.

海尚郡墅 • 锦华别墅

HAI SHANG JUN VILLA•JIN HUA VILLA

设 计 师：连自成
参与设计：刁秋蓉、袁晓凡、金李江
项目地点：上海
项目面积：712 m²
主要材料：米白洞石、珍珠鱼皮、铁刀木、珍珠贝母、意大利木纹石、橡木、霸王花大理石等
摄 影 师：张嗣叶

Designer: Lian Zicheng
Participatory Designers: Diao Qiurong, Yuan Xiaofan, Jin Lijiang
Project Location: Shanghai
Project Area: 712 m²
Major Materials: Beige travertine, Pearl skin, Iron knives wood, Pearl fritillaria, Italian wood stone, Oak, King flower marble, etc.
Photographer: Zhang Siye

本案在设计的时候，更多地运用具有张力的设计感去阐述人内在的需求，用空间表达内在的诉求。别墅的内部空间表达，附着于建筑架构之上，通过异型材质的搭配增加了空间的透视感和层次感。精简的新艺术的线条穿插于整个空间中，视觉点得到极大的延伸。新东方风格的装饰语汇糅合西方的空间架构，交互式的文化撞击打破了固有的设计风格的束缚，色调及木质基调的交互相映，减弱了冲突感。

地下室在设计构思上被隔断成若干的功能性区域，气氛更加轻松活跃。半开放式的隔断营造出呼吸通透的区间，拨动着和谐的韵律。松树、花、琴棋书画的古典东方陈设展现出鲜活的能量，并渗透于每个角落。

在采光上，设计巧妙地利用了地下室的天光。虚与实的表现，明与暗的交替，具有鲜明的层次感和立体感，使得室内外的环境融为一体。顺应自然光线的轨迹，空间赋予的能量延展到每个角落。

东方和西方的相遇，是文化的交叠，也是设计者和居住者的思维碰撞。设计传达的空间信息，是居住者生活态度的体现。

Photo-
wisdom

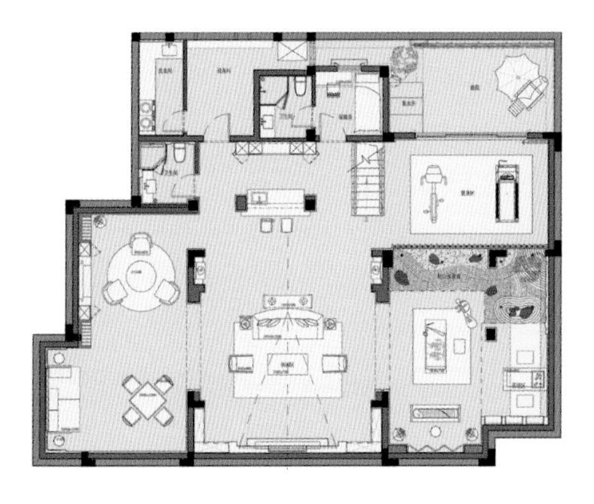

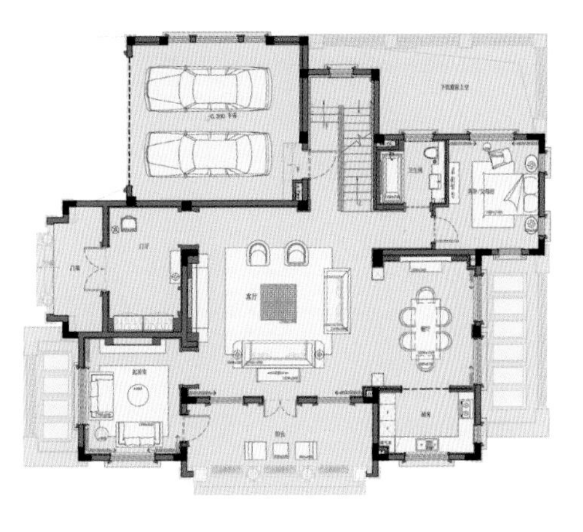

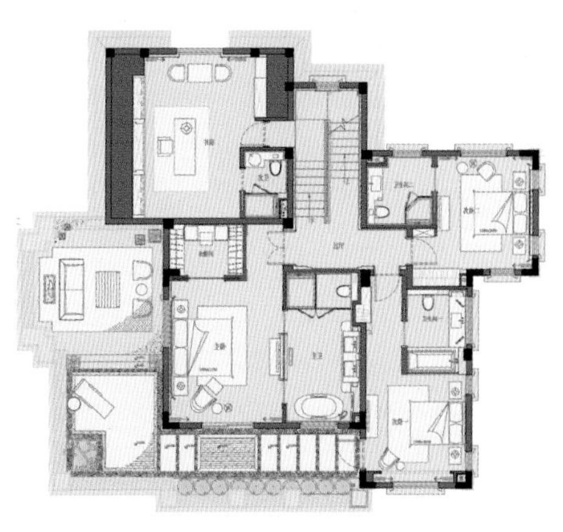

The case design applies extension sense to explain the inner need of people, using space to express the inner requests. The inner space expression is attached on the architectural construction. The match of different materials adds the space's transparency and layer sense. The simple new art lines insert through the whole space, and the visual effect gains large extension. The new-oriental style decoration combines with the western space construction. The exchange culture conflict breaks the confined design style. The echo of color tone and wooden background color releases the conflict sense.

The basement has been interrupted into several functional areas, to make the atmosphere more relax and active. The half-open partition creates a breath area, which is transparent and harmonious. The classic orient display of pines, flowers, pianos, chess, books, paints shows vivid power that penetrates in each corner.

On the use of light, the designer cleverly uses basement light. The fictional and real shadow has vivid layer sense and 3D sense. The inner and outdoor environment are combines together. The natural light in the space extends to every corner of the space.

The meeting of east and west is the folding of culture, and also the clash of designers' thinking and residents' thinking. The space information of design is an expression of residents' life attitude.

GI10 住宅案

GI10 RESIDENCE

设计单位：台北玄武设计
参与设计：黄书恒、欧阳毅、
陈佑如、张铧文
软装设计：胡春惠、张禾蒂、沈颖
项目面积：149.71 m²
主要材料：黑白根、镜面不锈钢、黑
蕾丝木皮、银箔、金箔、进口拼花马
赛克、黑白色钢烤
摄 影 师：赵志程
文案撰写：程歆淳

Design Company: Taibei Sherwood Design
Participatory Designers: Huang Shuheng, Ouyang Yi, Chen Youru, Zhang Huawen
Soft Decoration Designers: Hu Chunhui, Zhang Hedi, Shenying
Project Area: 149.71 m²
Major Materials: Black and white marble, Mirror stainless steel, Black lace veneer, Silver foil, Gold foil, Imported pattern mosaic, Black and white steel grill
Photographer: Zhao Zhicheng
Copywriter: Cheng Xinchun

GI10一案为坐落于城市新区的宅邸，有半山坡的绿意相伴，从客厅落地窗放眼望去，广场的宽大视野，也成为居所的重要亮点，作为退休生活的开始，必然需要一番缜密而细腻的规划。玄武设计考虑到业主姐弟与母亲同住的需求，以及居住者对于美学风格的爱好，力求艺术生活化，生活艺术化，最终择以现代巴洛克为基调，以其独有的收敛与狂放，配合玄武擅长的中西混搭冲突美学，铺陈空间每寸轴线。

尚未进入玄关，已见一座当代艺术作品灵动而立，既巧妙掩饰了半弧形缺角，又为居所带来了活跃的生机；右进，切入高耸柱式与圆形顶盖，视觉猛然挑高，使人豁然开朗，经典的黑白纯色打底，配合景泰蓝珐琅与定做琉璃，为访客带来第二重震撼。

业主因业务所需，时有交际与公务的需求，特别需要既大气又有情趣的客厅空间，方便与客人聊天。玄武设计注重天然风光与人为艺术的调和，保留大型落地窗与沙发的间距，后者特别选用进口原版设计，呈现简练利落的现代风情。中央大胆置入以艳紫、宝蓝与金黄三者交织而成的地毯，强化了简约与繁复的冲突美感。

抬眼向上，一盏华丽的银色花朵灿烂夺目，使人倍感震撼，这座取材苗族银饰的大型艺术品，为玄武设计与当代艺术家席时斌共同创作，外围用鸾尾花意象，曲折花饰包复核心，间隙镶嵌彩色琉璃，使打底的银灰色更显时尚。每当开关按下，艺术品外围即有五彩灯光流转，可因不同情境而切换，上缀羽饰的大型银饰环绕着核心缓缓移动，隐喻着天文学中恒星与行星的概念，呈现着自然与人文的灵动对话。

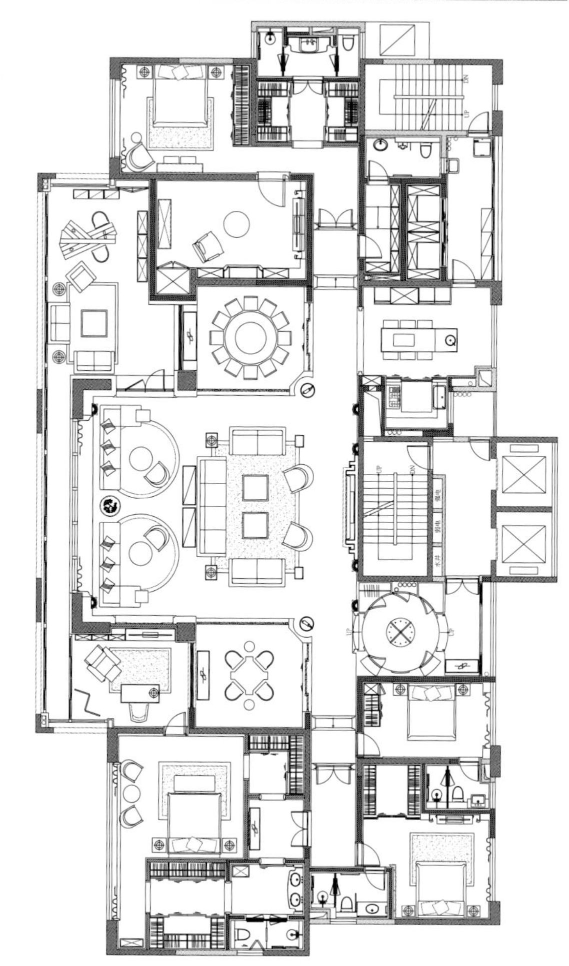

穿越廊道，可进入业主的阅读空间。主卧书房一方面延续着公共空间的半圆形语汇，引导访客进入皮质沙发、深色书柜、石材拼花共构的豪气场域，跳跃着清淡的色泽，大幅提升空间的律动感；第二主卧的书房，则纳入半户外的开阔设计，以白

梅花 王安石
墙角数枝梅，凌寒獨自开。
遥知不是雪，为有暗香来。

色铺底，使用黑色书柜与铁灰沙发，抢眼的小号造型灯具，充分体现了屋主的喜好。

公共区域的门扉使用白色，给人亲近、纯净之感；进入私密区则以黑色区隔；进入次要空间，棋牌室与餐厅分据左右，二者均以白色为主调。黑白格地板，置入经典款水晶灯，搭配巴洛克花纹座椅、鸽灰抱枕。

为使主客起卧舒适，主客卧房采用一贯的轻柔色泽，再以方向不同的线条勾勒空间表情。主卧简练的长形线板，与金黄床褥、浅蓝地毯相映成趣，减少过度堆砌的冗赘感；其余卧房则以湖水绿、天空蓝为点缀，在纯白、浅灰的基调里，窗帘、床褥与地毯稍有呈现，与牡丹纹床背板的繁复，共谱出业主悠闲淡雅的生活情趣。

GI10 case locates at the new city area. It is accompanied by mountain green. Seeing from the living room's french window, the wide vision of the square becomes the important light of residence. As the start of retire life, the house needs to be detailed planned. The Sherwood design takes into consideration of the living need of owners' sister and brother and their mother and the love of residence's aesthetic style of the residence. The aim of the design is to make life art and vice versa. The tone is modern Baroque style. The sole convergence and wild and Chinese and Western aesthetics of conflict is full of each axis in the space.

Before entering into the entrance, people can see a contemporary art work in the sight. It hide the half-arch corner cleverly, and brings the house active sense. Stepping into the house from the right way, there is a high column and round roof. The sight is suddenly becoming high, making people feel open. Classic black and white basement and cloisonne enamel with glass brings second shock to the visitor.

Because of the job, the owner always has to attend social life and business work. Therefore, they need a living room space which is interesting and free. It must is convenient to chat with the guests. Sherwood Design emphases on the coordination of nature scenery and human art, maintaining the distance between large french window and sofa. The sofa is imported, showing succinct and neat modern sense. The center of the house can be added with purple, blue and yellow carpet, enhancing simple and complex conflict of beauty.

Looking up, a light dazzling ornate silver flower makes people feel shocked. This large scale Miao silver artwork are designed by Sherwood design and contemporary artist Xi Shibin. Peripheral iris images are designed with tortuous flower and core in the center. Between the petal, there is inset with colorful glass, makes the bottom gray color more fashion. Each time the switch is pressed, the artwork is surrounded with multicolored lights. The light can be changed by different situation. The large silver products with

feather slowly moves around the core, standing for the astronomy—stars and planets' concept, showing a dialogue between Nature and Humanities.

Through the corridor, one can enter into the owner's reading room. The master bedroom on the one hand continues the semi-arc vocabulary of the public space, guiding visitors into the area containing leather sofa, dark bookcase, stone mosaic. The jumping light color, significantly increases the rhythm of the space. The study of the second main bedroom can be included in the semi-outdoor design. The bottom color is white, the bookcase is black and sofa is grey. The small eye-catching lamps fully reflect the owner's love.

The doors in the public area are white, giving people close and pure feeling. The private area uses color black. The secondary space contains chess room and dining room, both are white. The black and white floor has classic crystal lamps with baroque pattern seat and dove gray pillow.

In order to make the main bedroom smooth, the house uses gentle color and light color. And then designer adds lines in different spaces to sketch different spatial expression. The master bedroom's concise long strip, golden mattress, light blue carpet forms a delightful contrast, reducing excessive piling verbose sense. The remaining bedrooms are decorated with lake green, sky blue. The curtain, the bedding and carpet are in white, light gray tone. They and the complicated peony bed back panel compose a relaxed and elegant life together.

杭州丽景山别墅

HANGZHOU LI JING MOUNTAIN VILLA

设计单位：深圳市帝凯室内设计有限公司
设计师：徐树仁
项目地址：浙江杭州
项目面积：500 m²

Design Company: Shenzhen Dikai Interior Design Co., Ltd.
Designer: Xu Shuren
Project Location: Hangzhou, Zhejiang
Project Area: 500 m²

人们开始摒弃繁缛豪华的装修，力求拥有一种自然简约的居室空间。简约中式风格脱离传统中式的烦琐，少了中式的沉闷，多了温馨感和现代感。本案设计手法简洁，空间配色轻松自然，又能在简单的中式元素运用中体现出中国传统文化的魅力。简单的木色，精致的线条勾勒，大面积的白，沉稳大方，奢华，又不失品位……

People start to giving up luxury decoration, and seek for a kind of natural and simple space. Simple Chinese style separate itself from traditional Chinese style's complication. It is lack of Chinese style's dull while has more warm and modern sense. The case's design method is simple, the space color is relaxed. The apply of simple Chinese elements shows the charm of Chinese traditional culture. The simple wooden color, refined line sculpture and large white are fine, stable and full of tastes.

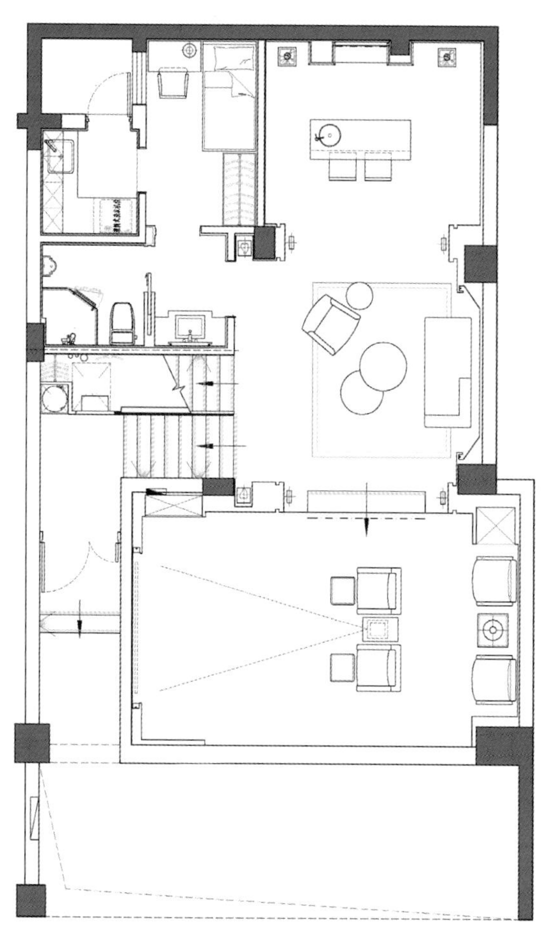

KELLY HOPPEN
KELLY HOPPEN
KELLY HOPPEN STYLE

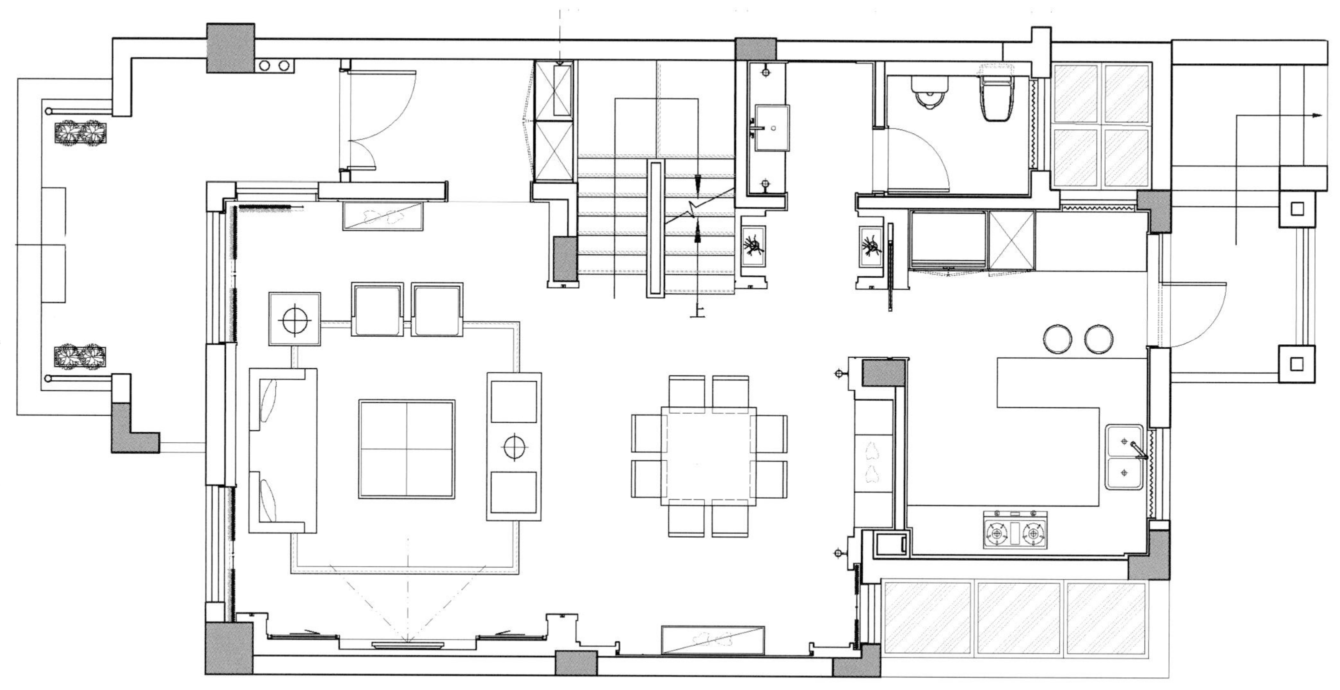

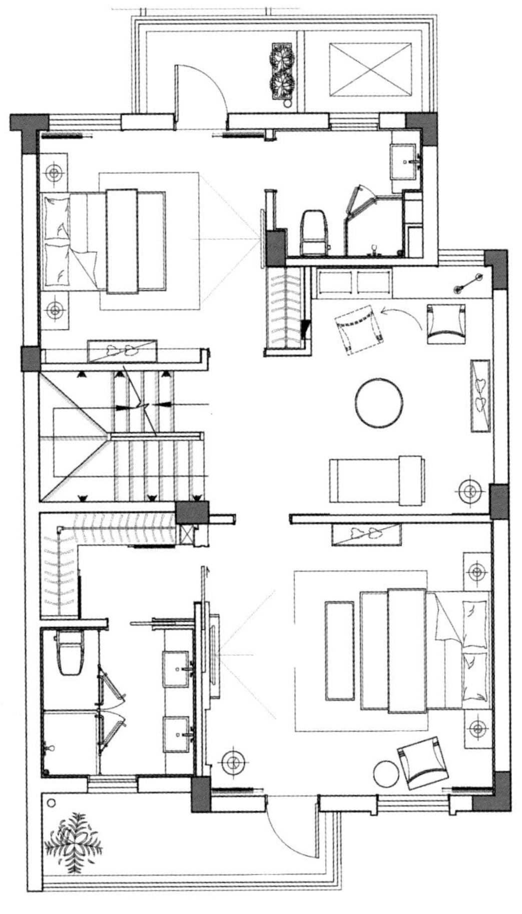

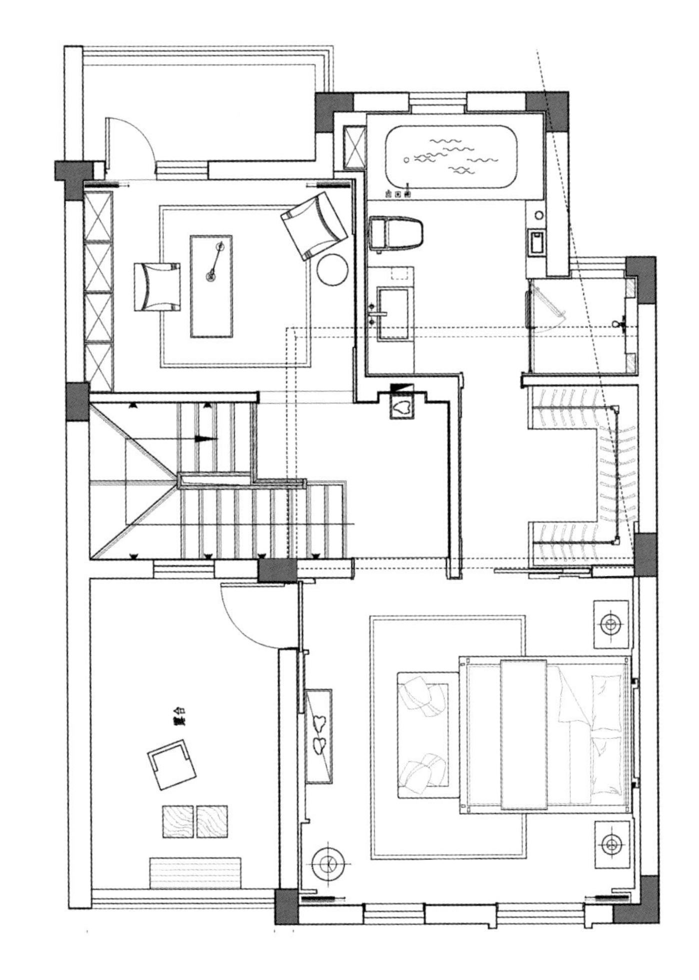

虹景花园私人别墅

HONGJING GARDEN PERSONAL VILLA

设 计 师：马梓滔
设计面积：600 m²
主要材料：天然大理石、马赛克、实木地板、实木定制墙板、进口壁纸

Designer: Ma Zitao
Design Area: 600 m²
Major Materials: Natural marble, Mosaics, Wood flooring, Solid wood custom wall plate, Imported wallpaper

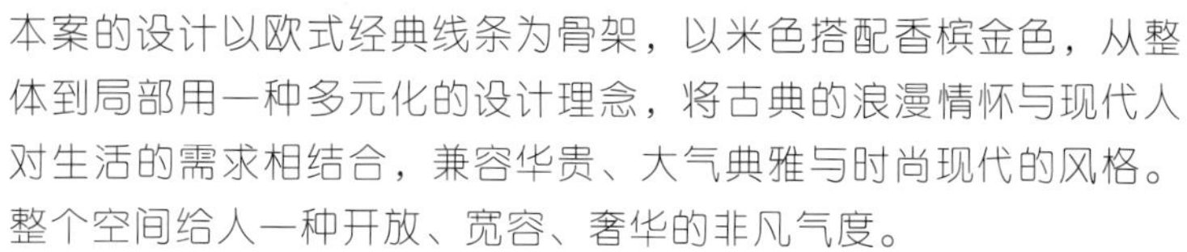

本案的设计以欧式经典线条为骨架，以米色搭配香槟金色，从整体到局部用一种多元化的设计理念，将古典的浪漫情怀与现代人对生活的需求相结合，兼容华贵、大气典雅与时尚现代的风格。整个空间给人一种开放、宽容、奢华的非凡气度。

The case design bases on European classic lines. The color is beige and champagne gold. It uses a diversified design concept from the whole to the part. Design combines classic romantic sense and modern people's needs, containing luxury, free, elegant and modern style. The whole space gives people open, tolerant and luxurious sense.

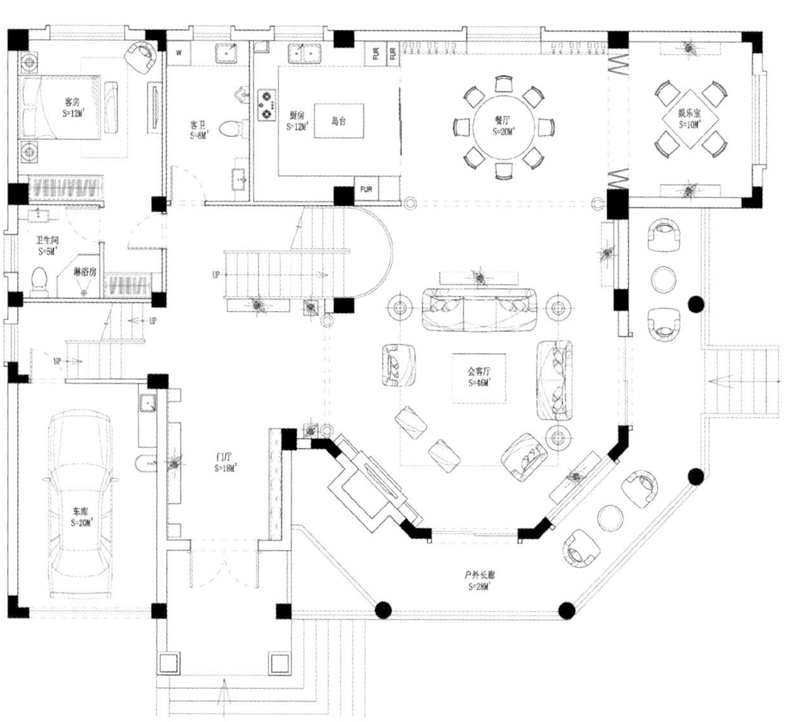
客房
S=12M²
客卫
S=8M²
厨房
S=12M²
岛台
餐厅
S=20M²
娱乐室
S=10M²
W
FUR
卫生间
S=5M²
淋浴房
UP
会客厅
S=46M²
门厅
S=18M²
车库
S=20M²
户外长廊
S=28M²

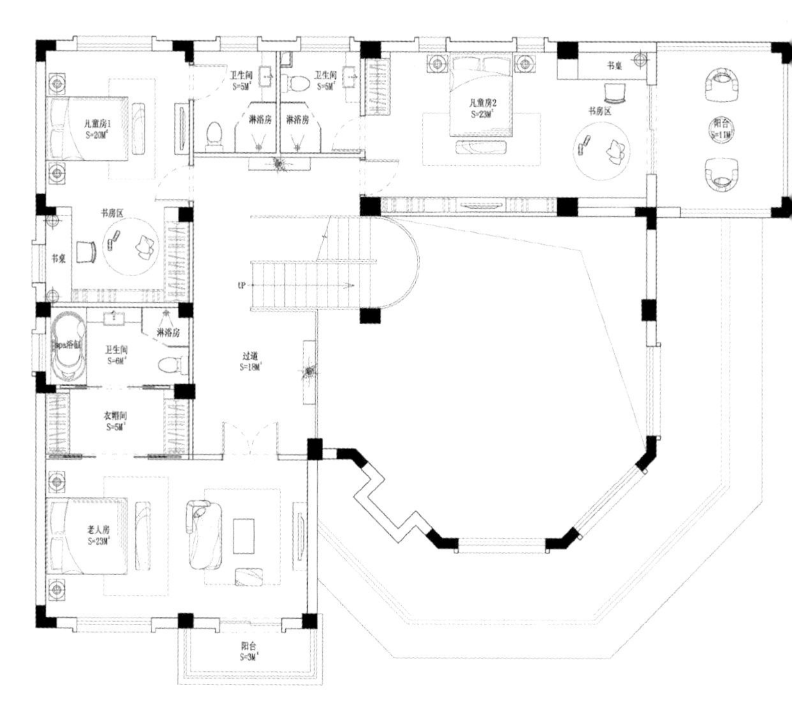
儿童房1
S=20M²
卫生间
S=5M²
卫生间
S=5M²
淋浴房
淋浴房
儿童房2
S=23M²
书桌
书房区
阳台
S=11M²
书房区
书桌
UP
spa浴缸
卫生间
S=6M²
淋浴房
过道
S=18M²
衣帽间
S=5M²
老人房
S=23M²
阳台
S=3M²

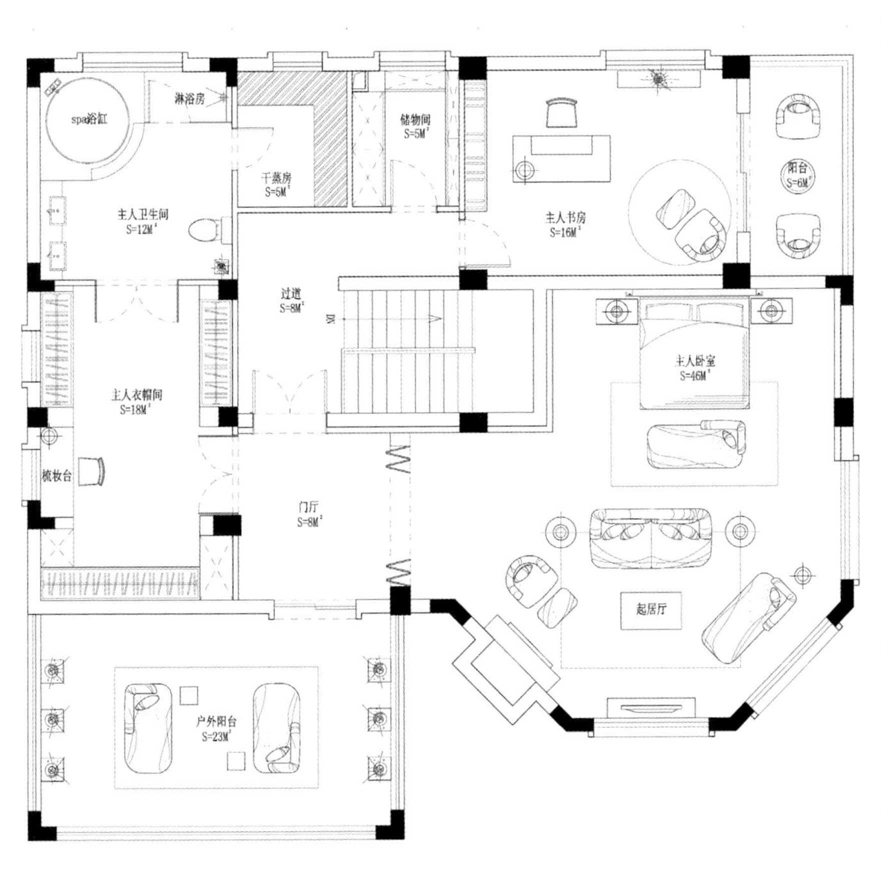
spa浴缸
淋浴房
储物间
S=5M²
干蒸房
S=5M²
主人卫生间
S=12M²
主人书房
S=16M²
阳台
S=6M²
过道
S=8M²
主人卧室
S=46M²
主人衣帽间
S=18M²
梳妆台
门厅
S=8M²
起居厅
户外阳台
S=23M²

柳心序

LIU XIN SEQUENCE

设计单位：花开设计师工作室
设计师：林函丹
项目地点：福建福州世贸天城
项目面积：200 m^2
摄影师：施凯

Design Company: Flower Blossom Designer Firm
Designer: Lin Handan
Project Location: World Trade City, Fuzhou, Fujian
Project Area: 200 m^2
Photographer: Shi Kai

事业有成的女人让人羡慕，她对生活有着美好的追求，喜欢旅游、下厨、看书。对自己的住宅装修要求也比较高。业主在找到花开设计之前曾找过别的设计师对装修进行设计，但对户型布局很不满意。原户型动、静区域分隔不明显，其儿子的房间又比较小，原卫生间容纳不下业主钟爱泡澡所用的浴缸。为了使业主满意，花开在户型设计上做出很大调整，大胆地把客厅与儿童房间做了重新设计，分隔成动、静区域。动区有中式厨房、敞开的西厨区、客厅及多功能的餐厅区；并将西式卡座式的餐厅改造为多功能区，变成用餐、朋友聊天品茶、上网、观看窗外风景的专用空间。原来的客厅阳台因为客厅改变后直接改变成主卧泡澡区和淋浴区。户型改造的图纸出来后，业主非常满意，施工期间一次都没来看过，我们施工团队用4个月时间保质保量地完成了装修工作。

People envy career woman. The owner is a career woman, she has happy pursuit towards life. She like traveling, cooking and reading. She has high requirements towards residential decoration. Before she find Blossom Design, she has looked for other designers to design the house, but she was not satisfied with the layout. The original type has no obvious static and dynamic region. The son room is small. The bathroom can not hold the owner's favorite bathtub. In order to make the owner satisfied, Blossom Design made a great adjustments in type. They boldly redesign the living room and kid room, separating the space into static and dynamic. The dynamic area has

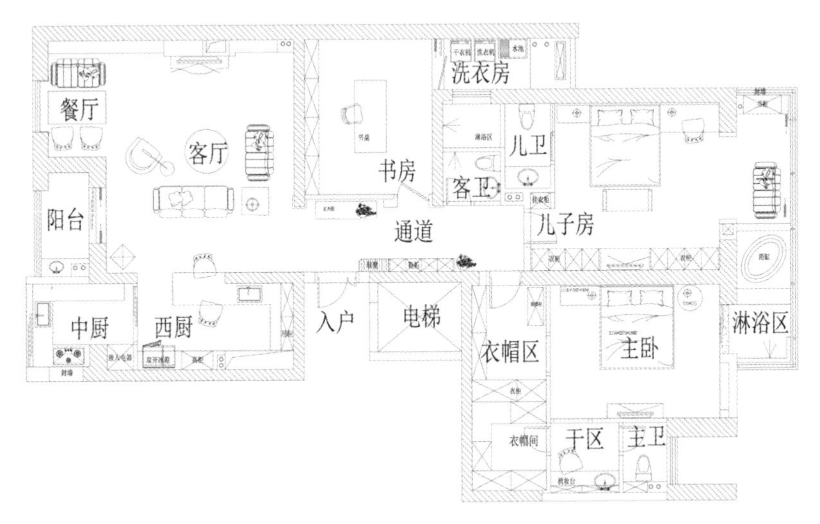
餐厅
客厅
书房
洗衣房
儿卫
客卫
儿子房
通道
阳台
中厨
西厨
入户
电梯
衣帽区
主卧
淋浴区
衣帽间
于区
主卫
梳妆台

Chinese-style open kitchen, Western open kitchen, living room and multipurpose dining room. The Western-style restaurant has been transformed into multifunctional area, people can eating, talking to friends, drinking tea, surfing the net, watching the scenery outside the window here. The original guests room's balcony can be directly used as shower area. The owner is very satisfied with the type transformation paper. Thus she is not coming to the construction area during the construction area. We use four weeks to completely finishes the decoration work.

诡境

GUI JING

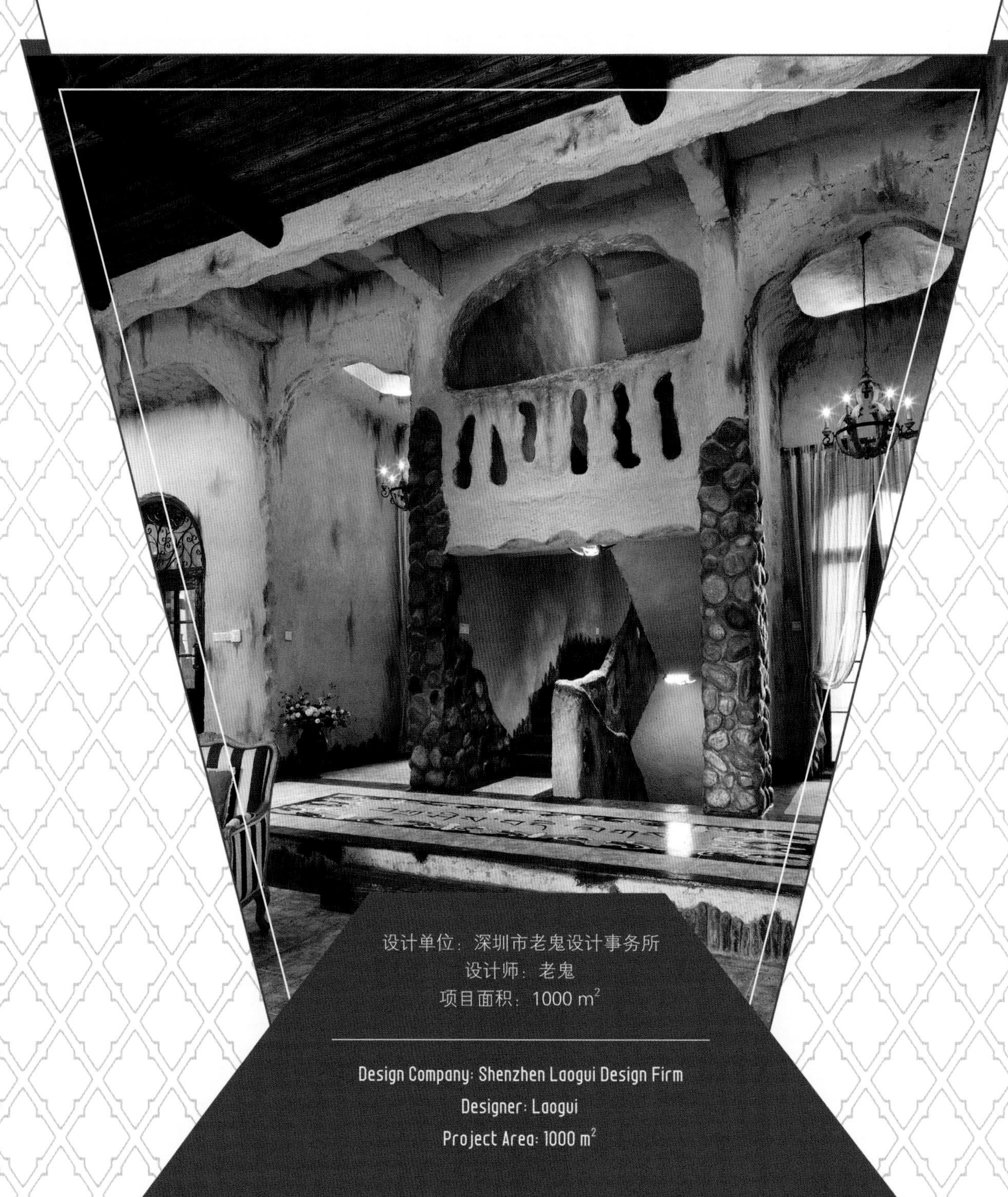

设计单位：深圳市老鬼设计事务所
设计师：老鬼
项目面积：1000 m²

Design Company: Shenzhen Laogui Design Firm
Designer: Laogui
Project Area: 1000 m²

当别人问起“您这是什么风格？”时，“鬼派设计”是最好的诠释。事实上，在设计之初，老鬼并不想要一种被大多数人贴上的风格标签。但是老鬼的作品依旧是“有迹可循”的。自幼学习书画的老鬼，将书画艺术中的写意与意象之美融入到了设计之中，再加上其特立独行的“鬼派”设计语言，最终实现的空间效果既有着人间的舒适，又有着“诡境”的灵动气韵。摒弃所谓高大上的设计元素，从生活中选材。打破习惯的规矩，从自然中寻找规则。河边捡来的石头、地上扒来的砖块、被人遗弃的破损石像……皆可成为老鬼的设计元素。这种“任性”是对自然的尊崇，也是对自己的肯定。他不去试图改变自然之美，而是在自然之美上加上变幻莫测的“鬼计”，使之“气韵生动”。

THANK
YOU
THANK
YOU

When someone else asks, “What kind of style your design belongs to?” , the best answer is that “It is a ghost style.”. In fact, at the beginning of the design, Laogui is not like to be pasted with a piece of fixed-style label. While Laogui’s work follows a kind of “principle”. Laogui has been studying painting and calligraphy since he was young. He applies the beauty of image into his design work, with his unique “ghost style design language”, as a whole, they form a smooth human-world space with “fantasy feel”. The design throws away the high-end design elements, choosing materials from life. It breaks the old custom rules, finding new principle from the nature. Stones at the river bank, bricks on the ground, useless broken statues...they are all Laogui’s design elements. This kind of “self-indulgence” is a respect towards nature as well as a self-acceptance. Laogui does not try to change the nature beauty, but adds capricious “ghost feel” on it, making it full of “artistic conception”.

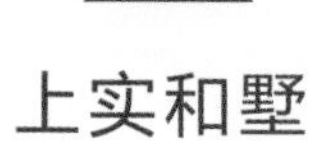

上实和墅

SHANG SHI HE SHU

设计单位：上海鼎族室内装饰设计有限公司
设 计 师：吴军宏
项目地点：上海
项目面积：550 m²
摄 影 师：上海三像摄文化传播有限公司 张静

Design Company: Shanghai Dingzu Interior Decoration Design Co., Ltd.
Designer: Wu Junhong
Project Location: Shanghai
Project Area: 550 m²
Photographer: Shanghai San Xiang She Culture Promotion Co., Ltd. Zhang Jing

新拓明代绘画精品集
龍藏寺碑

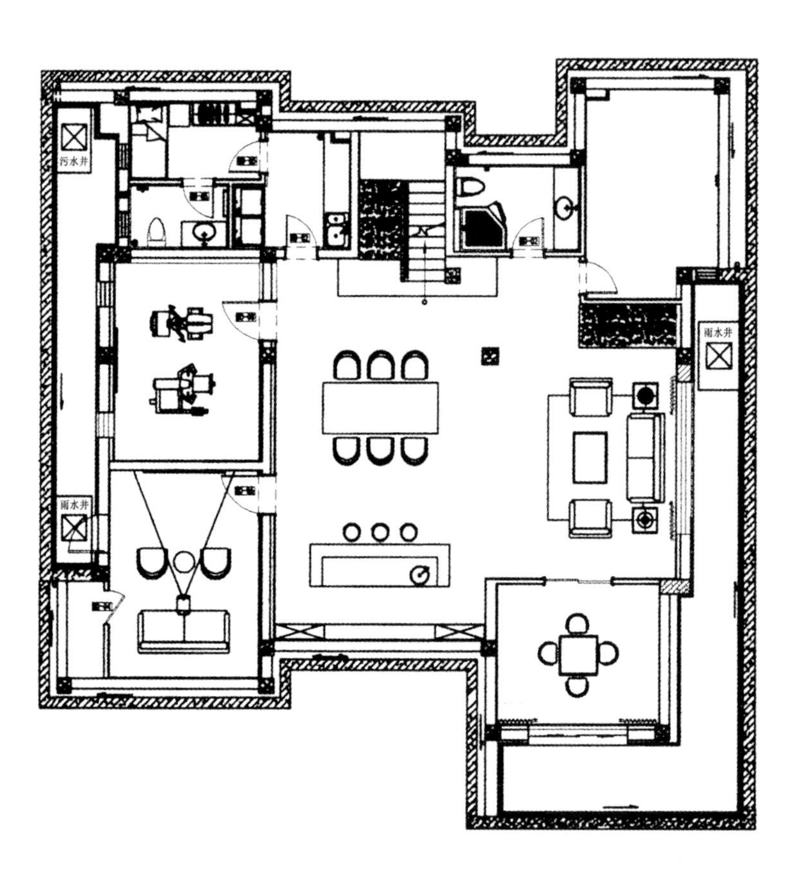
污水井
雨水井
雨水井

低碳是当下建筑装饰行业为众人所瞩目的课题，而由于木材在生长过程中固化的碳，比被砍伐、生产和使用过程中释放得更多，因此它在某种意义上实际是一种“负碳”建材，于是，木结构住宅又被称为“会呼吸的房子”。和墅被称为是在用全世界最好的原材料——加拿大的木材，全世界最先进、最成熟的经验——日本的技术，建造最适宜中国市场、引领未来宜居方向的木结构豪宅。

设计师在考量本案的时候，充分重视了和墅的环保独墅理念和对传统文化的认知。本案装饰中，设计师运用了陶瓷质地的花瓶和茶具、文房四宝、国画、书籍、几案、书桌、官帽椅、有着宁式百工床风格的卧榻等等，共同构成了有机的中式元素组合体系。室内对木饰面的大量运用，并没有刻意营造一种西式宫廷的奢华，而是和明式家具一起构成了传统中式风情的格调。

我们可以设想，在茶香浓郁的茶室，悠然品茗的是独具慧眼的业主；在书香飘逸的书房，“刚日读经，柔日读史”的也是这样的业主；而在花气袭人的卧室，安然入梦的更是这样的业主。在中式的传统中，居室从来不单单是纯粹的硬件，而是可以在其中“讲书肄业，琴歌酒赋”的心灵皈依之所。

中式的设计风格对设计师无疑是一种挑战，一个歧途，很容易过度使用相关元素，本意是追求富贵之气，却弄巧成拙，反而出现事与愿违的土财主式的堆砌，这种对格调的伤害是致命的，从这个角度，敢于使用中式风格的设计师可以说是具有充分的艺术自信，本案就是这样的一个典范。碧山人来、可人如玉的清奇在这里融汇在了一起，共同书写着关于一个会呼吸有情怀的住宅的传奇。

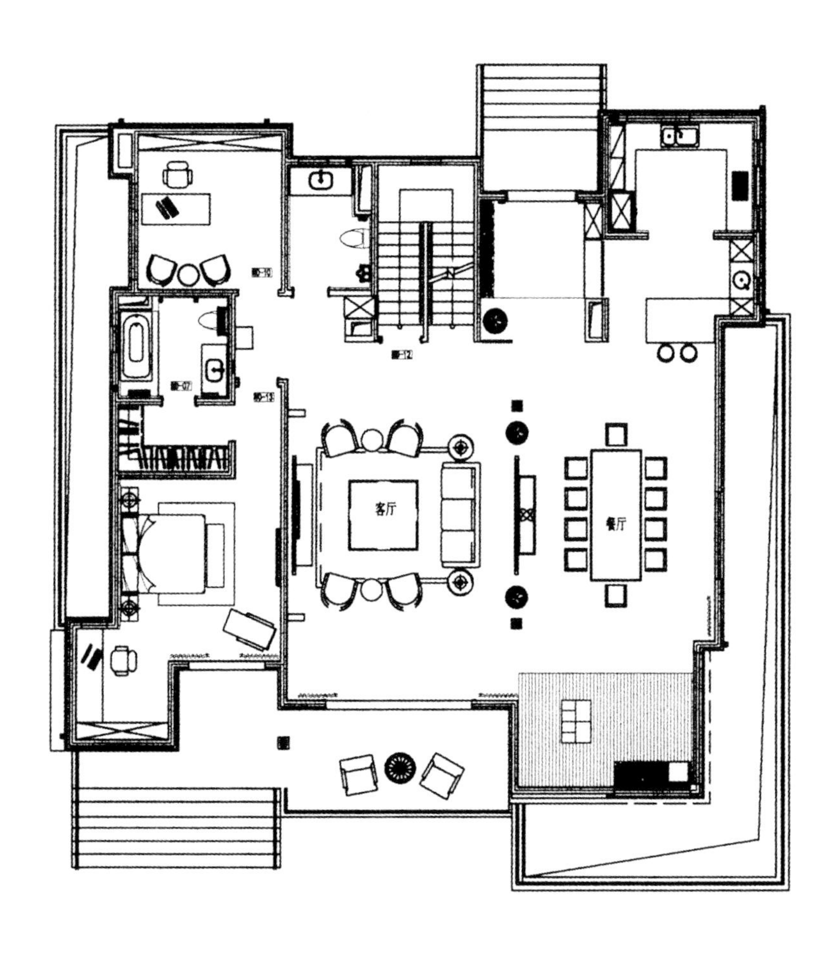
客厅
餐厅

Low-carbon is a famous project in the current architectural decoration industry. Since the carbon in the grow process of the wood is more than that reproducing in the process of harvest, production and use. To some extent it is a "carbon-bearing" material. The wood structure house also be called as "breathing house". He Shu is using the world's best raw material ---- Canadian wood, the world's most advanced, most sophisticated experience — Japanese technology, to build the wood structure villa that is proper in China and leading future living orient.

When the designer considers the case, he makes full consideration of He Shu's environmental villa idea and of traditional culture recognition. In the decoration of the case, designers use ceramic vases and tea sets, four treasures, paintings, books, tables, book desks, Armchairs, and the couch of Ning bed style, etc., together constituting the organic combination of Chinese elements system. The large use of wood finishes in the interior house did not deliberately create a Western-style palace luxury, but forms a traditional Chinese style with Ming furniture.

We can imagine the scenery that the owner sits in the teahouse and drinking tea leisurely, reading books and histories in the

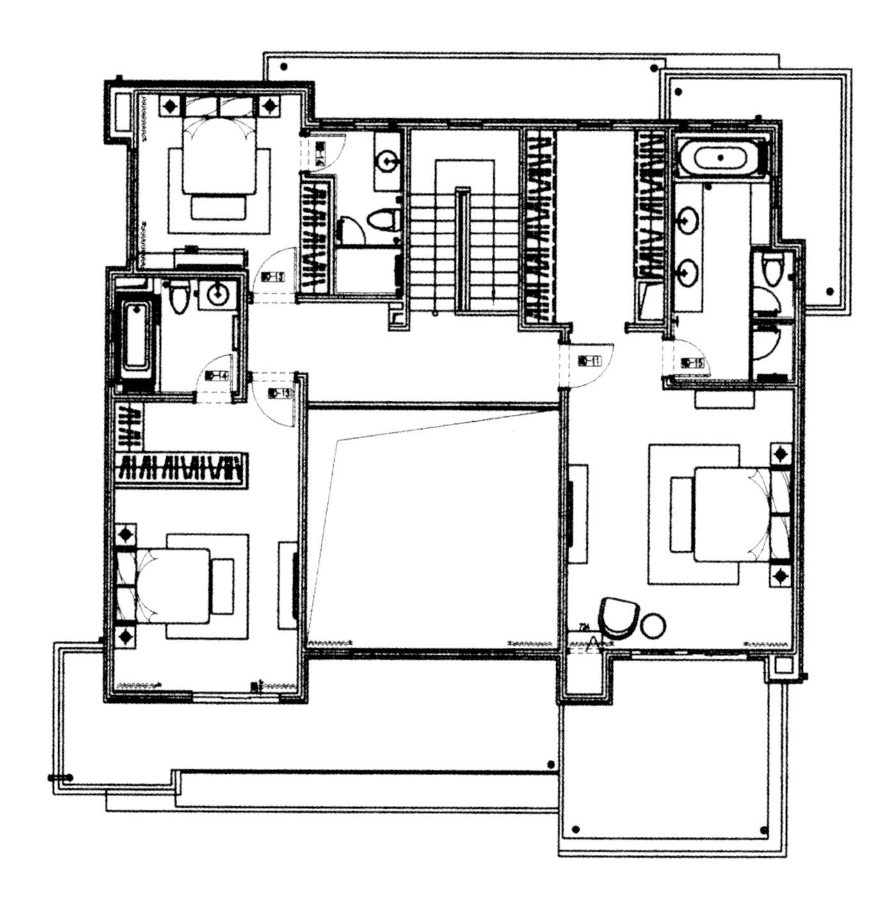

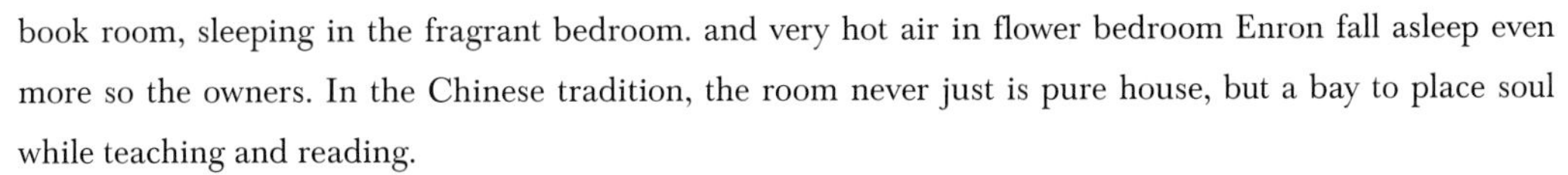

book room, sleeping in the fragrant bedroom. and very hot air in flower bedroom Enron fall asleep even more so the owners. In the Chinese tradition, the room never just is pure house, but a bay to place soul while teaching and reading.

Chinese design style is undoubtedly a challenge for the designers. The overuse of related elements may lead to inverse outcome. The rich fortune may seems like a stack of local wealthy. The damage of style is fatal. From this perspective, the designer dear to use Chinese-style can be said of having sufficient art confidence. The case is a typical example. The green mountain, jade-like bright and marvellous sense forms together in the house, writing a legend of a breath and emotional house.

安泰别业

AN TAI BIE YE

设计单位：大墅尚品
施工单位：大墅施工
软装设计：翁布里亚软装机构
项目面积：400 m^2
主要材料：地面为大部分为奥特曼理石、黑白根理石、爵士白理石
墙面为莎安娜理石、黑金花理石、花梨木面板、镜面玫瑰金
吊顶为白色乳胶漆

Design Company: Dashu Shangpin
Construction Company: Dashu Construction
Soft Decoration Design: Umbria Soft Decoration Institution
Project Area: 400 m^2
Major Materials: Floor: mainly are Altman marble, Black and White marble, Jazz White marble
Wall: Shaiana marble, Black and Gold Flower marble, Rosewood Panels, Mirror Rose Gold
Ceiling: White Latex Paint

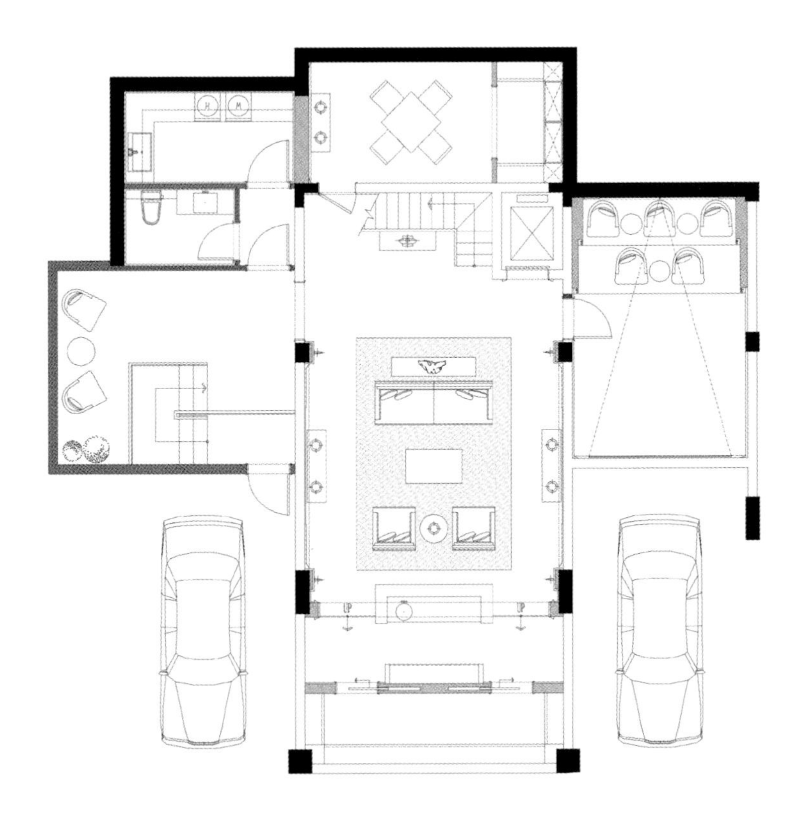

“家应该可以与主人一起成长。”家在不同时期可以有不同的呈现，展现不同的生活情怀。为此，这套住宅设计以“享生活”为主题创意概念，让家的空间从装修到家具，从装饰到陈列，无不展现出“生活”的浓厚气息。打破了空间固有设计风格的束缚，让空间可以随着生活阅历的积淀而呈现不同的变化。

整个空间以沉稳的木色为主基调，驼色为辅，空间环境雅致、奢华。宁静的装饰线条，时而闪烁耀眼，时而隐藏其型，或灵或透，一切都是最自然的呈现，让空间弥漫高贵的古典韵味。

“Family should grow together with the owner.” Family may has different appearance at different period, showing different life feelings. Thus, the house can be designed in the theme of “life enjoyment”. The family from decoration to furniture and display expresses “life”'s deep breath. Designer breaks the original design style's constraint. The space can change with the increase of life experience.

The space is based on stable wooden color, with brown color as supplement. The space is elegant and luxury. The peaceful decoration line sometimes is bright, sometimes is subtle. The natural transparency has noble classic charm.

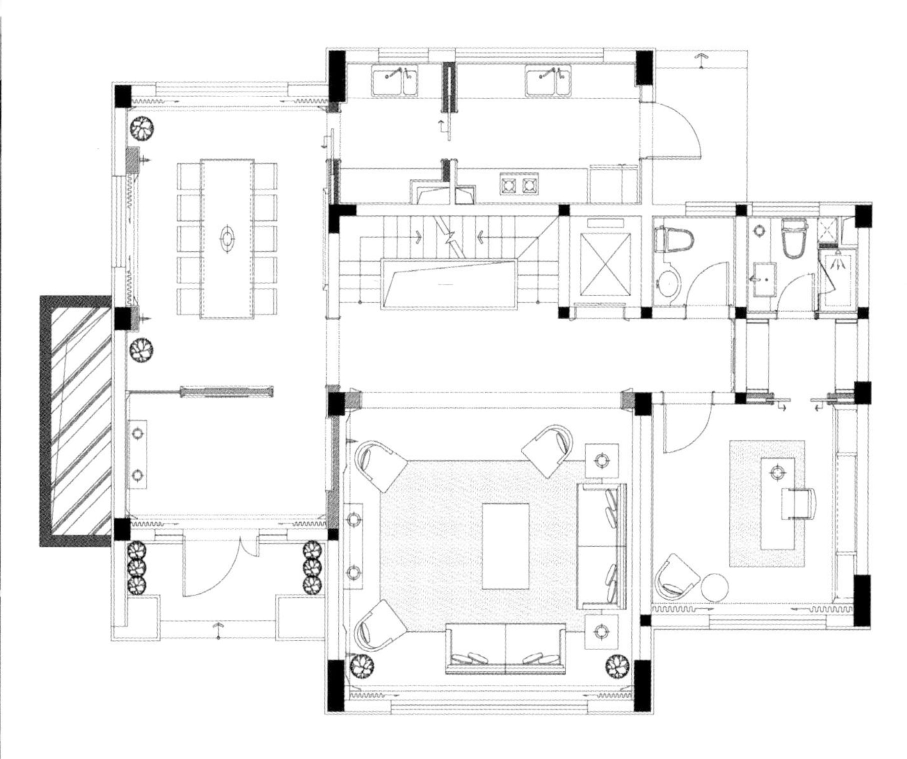

GUCCI
GUCCI

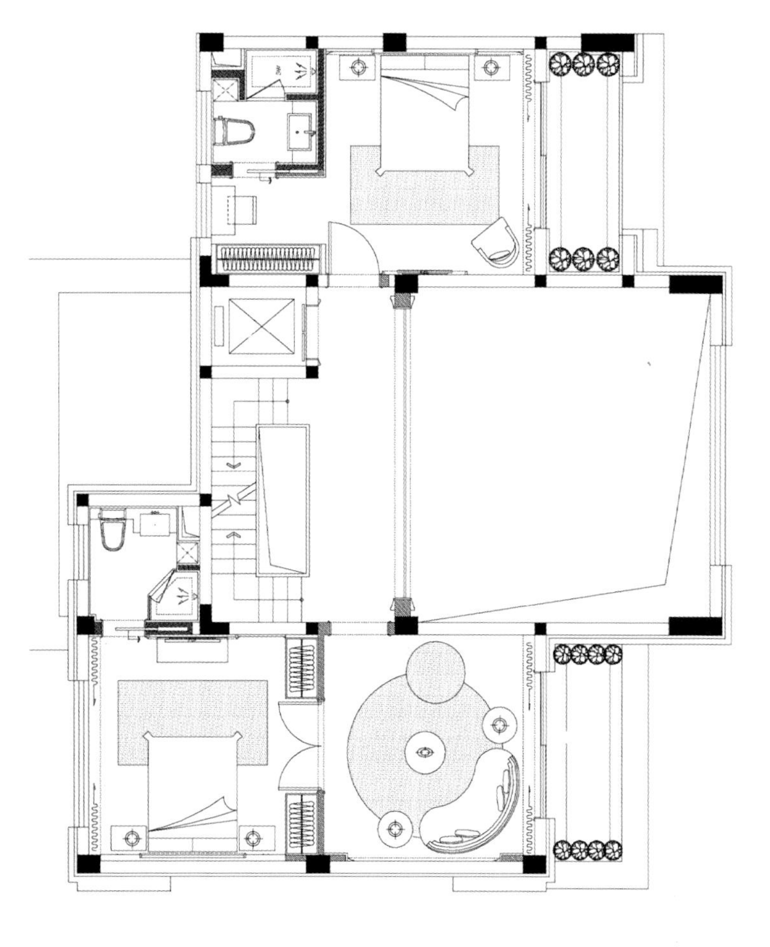

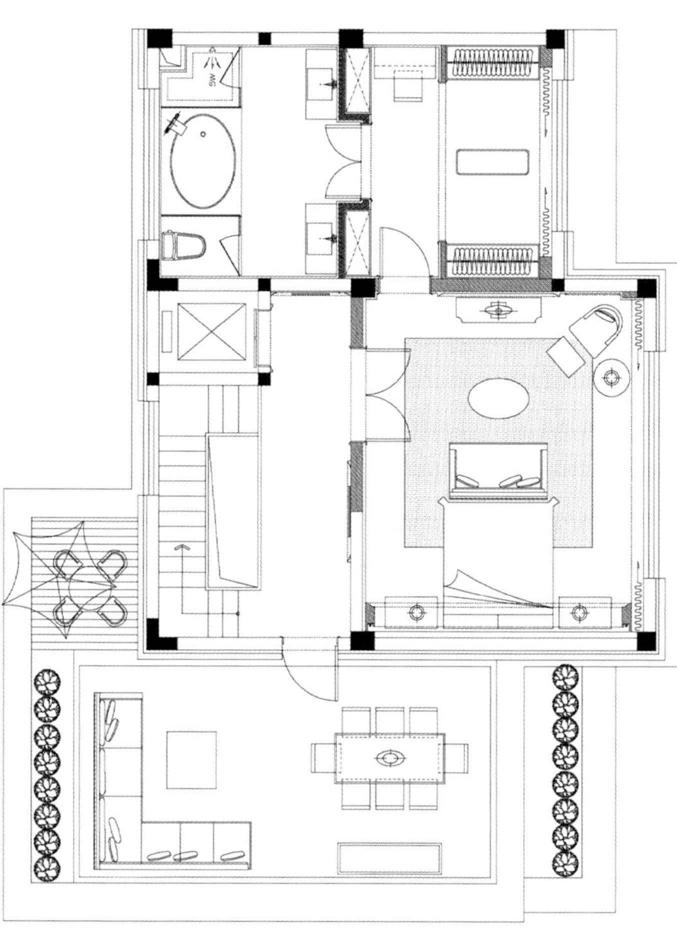

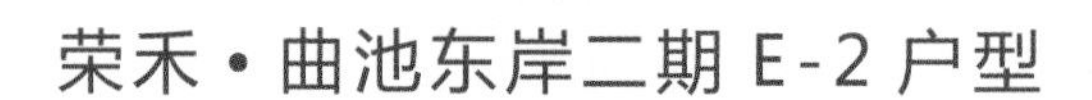

荣禾 • 曲池东岸二期 E-2 户型

RONG HE•QUCHI EAST LAND SECOND PHASE E-2 UNIT

设 计 师：郑树芬
参与设计：杜恒、黄永京
项目地点：陕西西安
项目面积：302 m²

Designer: Simon Chong
Participatory Designers: Amy Du, Jimmy Huang
Project Location: Xi' an, Shanxi
Project Area: 302 m²

“在这个世界里，轻歌曼舞尽日不息，声色犬马终年无休。萨克斯管彻夜吹奏着如泣如诉的‘华尔街蓝调’，上百双金色、银色的舞鞋踢起闪亮的灰尘。到了茶歇时间，这首低沉而甜蜜的热门歌曲依旧不断地回荡着，而许多新鲜的面孔宛如被那些铜管吹落在地面的玫瑰花瓣，在舞厅里到处飘来飘去。”菲茨杰拉德笔下的黄金年代是魅惑、鲜活、纵情享乐的纽约大都会时代。

独一无二的大都会气质，仿佛伦敦的优雅、米兰的时尚、巴黎的浪漫、东京的奢华、曼哈顿的大气、西班牙的激情，犹如一位成熟而有魅力的绅士细细品味着摩卡咖啡，静静地思索，悠远流长。

大都会的环境满足居住者的生活所需，开阔的格局因势而生，在基础空间布局上，开拓酒窖储藏室、酒吧区、园艺室及儿童娱乐室，将居者的生活兴趣融于居所。居住不再是居所单一的功能，社交、玩乐、休闲等高端私人会所的功能亦涵盖在内，引领上层生活需求的特性。

流行色“兰花紫”穿插在多个空间，使整个空间形成一个统一的整体，好比几幅相互关联、相互影响的系列画，你中有我，我中有你，又独树一帜，有着自己的特色。色彩鲜艳、抽象夸张的艺术挂画是这个空间大胆、刺激的设计元素。正如美国画家罗伯特•马瑟韦尔所说的：“艺术远没有生活重要，但是没有艺术，生活是多么乏味呀！”

"In this world, dance and sensual pleasures are all year round without a break. Saxophone playing throughout the night and hypnotizing listeners by 'Wall Street Blues', hundreds pair of shoes kicked up gold and silver shiny dust. At coffee break time, this low, sweet hot song is still constantly echoing many times. Many fresh faces are like rose petals felling on the ground , fluttering around the ballroom." The Golden age under Fitzgerald's pen is sexy, fresh, funny New York Metropolitan era.

Unique Metropolitan temperament is like London's elegance,

Milan's fashion, Paris's romance, Tokyo's luxury, Manhattan's freedom, Spain's passion; it is like a mature and charming gentleman savoring the Mocha coffee, thinking silently.

Metropolitan environment satisfies the needs of the residence, open pattern is coming out. According to basic space layout, design adds cellar storage room, bar area, garden room and children's play room, combining life interests into the house. Living is no longer a single function of housing, it also includes social function, fun function, leisure function and other high-end private clubs' function, leading the characteristics of upper life demand .

Popular color "purple orchids" intersperses in multiple spaces, so that the whole space forms a unified room, which is like several interlinked paintings. The paints are linking with each other, with special own characteristics. The colorful, abstract art paintings are the space's bold and exciting design elements, just like the United States painter Robert Motherwell said, "Art is much less important than life, but if there is no art, life will be so boring!"

荆溪福院十二园私人住宅

JINGXI HAPPINESS TWELVE GARDEN PERSONAL RESIDENCE

设计单位：深圳埂上设计事务所
设 计 师：文志刚、李良超
项目面积：380 m²
项目地点：江苏常州
摄 影 师：黄缅贵

Design Company: Shenzhen Gengshang Design Office
Designers: Wen Zhigang, Li Liangchao
Project Area: 380 m²
Project Location:Changzhou,Jiangsu
Photographer: Huang Miangui

该案位于江苏省常州市青果巷荆溪福院十二园，是荆溪置业倾力打造的徽派韵味传世别墅。业主在国外生活多年回国居住，业主希望既有国外的生活又不失中国的本土文化。本案以怀旧复古的方式贯穿始终，将东方与西方文化融为一体，把唯美与精致、自然与恬静、优美与儒雅演绎得淋漓尽致。指尖触及的每一个角落都能感受到居者的文化感、贵气感、自在感与情调感。

地下室主要为主人的休闲空间，色彩赋予空间强大的生命力，画龙点睛地增强了视觉层次感与冲击力，彰显浪漫、恬静的韵味；新古典元素，贵气的紫色布艺、盆栽绿植、明清元素的现代油画，让整个空间多了些轻松活泼和娱乐玩味之感。

设计借鉴江南园林的因景互借、移步换景的手法，就像这白色流转楼梯，原色实木台阶，简约变化的线条，晕黄的灯光，带着北欧淡淡的童话书香。它将这个家的所有居室与生活连接在一起，如静立在室内的艺术品，蜿蜒着迎接每一位步入者，又舒展着通向另一个空间。

从地下室上来对着的是整个房子的中心——中庭，中庭的南面是客厅，北面是餐厅及西厨厨房。原建筑是一个内庭院，只有1层高，我们把它加高到2层高，顶棚用玻璃封起来，让整个一层是一个全通的空间。黑色千层石打造的流水墙、木雕摆件等装饰品，如水墨画一般。

客厅大气内敛，儒雅含蓄。中西家具的混搭，彰显优雅不俗的装饰效果。颜色淡雅，释放现代之感。复古铜壶中的一支殷红梅花，让清雅中多了几分端庄丰华。

餐厅与户外的植物，盎然惬意，互为风景，诗韵潺潺。贵绿色和手工刺绣映衬得十分惬意，在午后阳光的照耀下，仿佛穿越时空，在记忆的故事里徜徉。一面是天井门庭的写意画，一面是窗外绿叶茵茵的自然之景，落座桌前，一餐一羹都显得新鲜可口，心意盎然了，便食之有味了。

主卧是风格鲜明的家具，更似一件工艺品展现出优雅不俗的装饰效果。以美式仿古的姿态，安于一隅的衣柜，在日月更迭、光与影的交替中，唤醒居者美好的记忆，释放向往自然、回归自然的情怀。天然白色洗石子的浴室，传统手工艺的质感，古朴的颜色，光影变幻，更具灵气与自然之力，与复古的居室相得益彰，娴静温柔地带来一种安全感。

一席阳光透过竹帘，婆娑地洒在室内，无需过多的饰品，矫揉造作的修饰，晨午黄昏，斑驳光影中，书房每时每刻都呈现不同的格调。云归处，人归来，华灯初上，几榻有度，位置有定，通透有序，含蓄不事张扬。无需焚香，新中式风格的古朴家具中，就透出幽幽的墨香。

作家梭罗曾说过："我们应该像攀摘一朵花那样，以温柔、优雅的态度生活。"好的居所，既没有富丽堂皇的气派，也没有毫无节制的排场，而是如此这般，从容有度，浮生优雅。

The case is located at twelve garden, Jingxi Fuyuan, Qingguo Xiang Changzhou City, Jiangsu Province, it is a Hui School villa created by Jingxi Zhiye. Owners are living many years in a foreign country and now they comes back to China. Owners hope that the house both has foreign style and Chinese local culture. The case uses retro way, combines Eastern and Western culture together, and combines exquisiteness and beauty, nature and tranquility together. Fingertip can feel culture sense, noble, nature and comfortable sense of every corner.

The basement is mainly the owner's recreational space. Color gives the space a strong vitality, enhancing a sense of visual hierarchy and impact, highlighting the romantic, quiet charm; neo-classical elements, noble purple fabrics, potted plants, the modern Ming and Qing paintings, making the whole space a lively and entertaining sense.

Design copies the techniques of Jiang Nan gardens. The white transition stairs, solid wood steps, minimalist lines, yellow light, and Nordic fairy tale books linking the home's bedroom with life, it is like the interior of art, windly greeting every people, and spreading to another space.

Coming up from the basement one can see the center of the House—the Atrium. Atrium's south part is the living room, north part is the kitchen and restaurant. The original building is a courtyard, only has one-storey high, we raise it to two-storey height, close the glass roof, thus the entire layer is a full space. Black layer stone's water walls, carved wooden ornaments are like ink paints.

Living room is free and elegant. The mixture of Chinese and Western furniture shows elegant and good decorative effect. The color is elegant, releasing modern sense. A red plum in a vintage copper pot, adding elegance and modesty in the fresh air.

Restaurant and outdoor plants are cozy and lush. Color green and hand-made embroidery are very nice. In the afternoon sun shine, people may feel across time and space, wandering in memory of the story. One side is the patio door painting, the other side is the moss natural scenery outside the

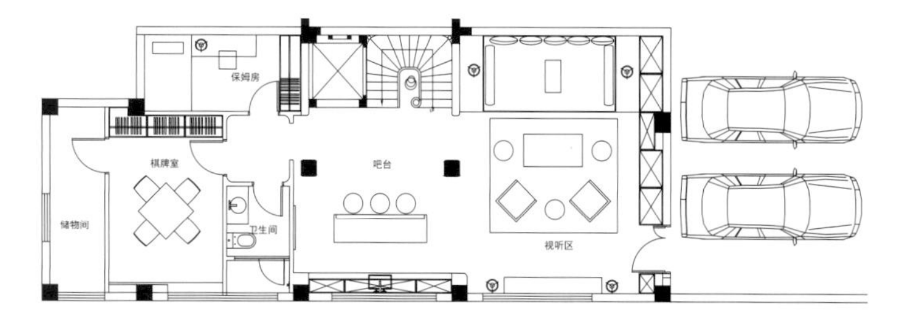
保姆房
棋牌室
吧台
储物间
卫生间
视听区

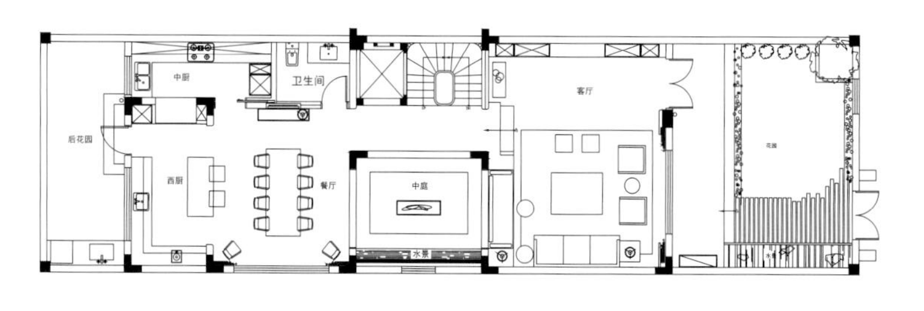
中厨
卫生间
客厅
后花园
花园
西厨
餐厅
中庭
水景

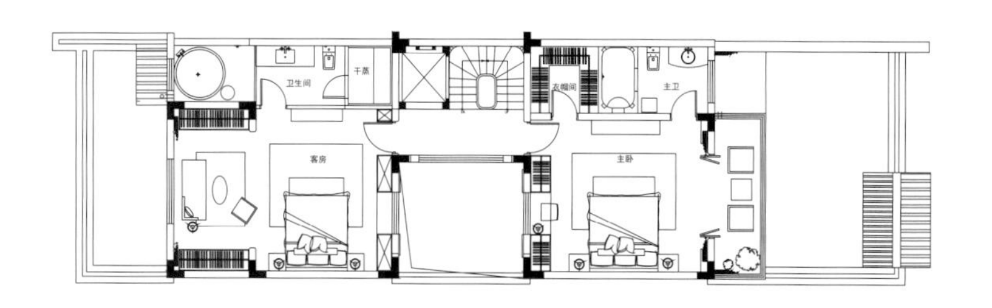
卫生间
干蒸
衣帽间
主卫
客房
主卧

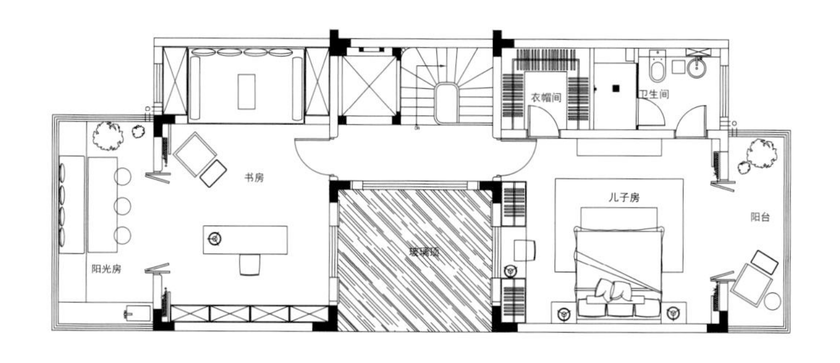
衣帽间
卫生间
书房
儿子房
阳台
玻璃顶
阳光房

window. Sitting in front of the table, the meal and soup are fresh and delicious, and the mind is happy.

The master bedroom has many distinctive style furniture, which is like an elegant art work with decoration effect. The designer uses American antique gesture, the closet in the corner wakes up the good memories in the exchange of sun and moon shadow, releasing the wish of going back to nature. The natural white washed stone bathroom, and traditional arts and crafts texture, plain colors, and the light change is full of brilliant and natural power. It echoes with the old house, bringing a sense of tender.

The sunshine lights through the bamboo curtain, showers into the interior room without excessive jewelry or artificial modification. In the morning, afternoon and evening light, the study shows different styles at every moment. The clouds return, people return, the lights are light, the location is stable, the space is transparent and order and very implicit. The new Chinese-style furniture has ink smell.

Writer Henry David Thoreau once said: "We should live with gentle, graceful attitude like picking a flower." Good accommodation is not palatial, and it has no unrestrained ostentation, and it is calm and elegant.

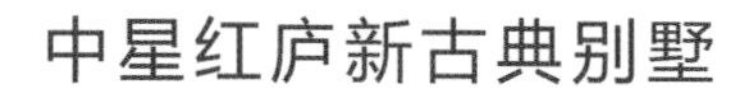

中星红庐新古典别墅

ZHONGXING HONGLU NEW CLASSIC VILLA

设计单位：上海鼎族室内装饰设计有限公司
设 计 师：吴军宏
项目地点：上海
项目面积：658 m²
摄 影 师：上海三像摄文化传播有限公司 张静

Design Company: Shanghai Dingzu Interior Decoration Design Co., Ltd.
Designer: Wu Junhong
Project Location: Shanghai
Project Area: 658 m²
Photographer: Shanghai San Xiang She Culture Promotion Co., Ltd. Zhang Jing

新古典的要旨就是用现代的手法和材质还原古典精神，同时具备古典与现代的双重审美效果，古与今完美的结合让人们在享受物质文明的同时得到了精神上的慰藉。本案的设计师用新古典来诠释这一具有人文精神的传承，无疑造就了一个经典脉络的延续和更新，处处闪现着文化的底蕴和魅力。

本案是古典和时尚的碰撞和相遇，爱奥尼克式的罗马柱、体现西方天人感应精神的堆砌壁炉、讴歌光明和爱的银质感五枝烛台、大面积的米黄和网纹大理石，以及金属质感的边桌等家具、抽象的装饰画、范思哲品牌元素，看似天南地北、上下千年，在这里，却形散神聚，完美地融合在了一起，彰显了业主作为拥有国际化视野的社会精英人士的眼光和品位。

步入客厅，屋顶的装饰处理和皮质沙发、成熟硬朗的灰色地毯透露出美式风格的自由和随性；餐厅的六人桌是西式大菜的标配，也呈现出浓浓的家庭氛围，墙上整体的镜面组合实现了空间感的延伸；几个卧室，设计师刻意用不同的或褐色或蓝色的色调区隔，形成了不同色系、不同色调的温馨感觉；厨房和卫生间的主体基调是现代的风格，一些新派的厨房设备被内置在整体厨房设施之中，低调而具功能性，卫浴设备也引入了罗马浴室的概念，闪烁着古典的浪漫气息。

特别值得一提的是书房、品酒区间和桌球室。本案的书房简洁明快，身为商业领袖的业主可以在这里神游八极、运筹帷幄；而在品酒区，无论是独酌还是与三五知己共饮，都能在微醺中放松身心，洗去红

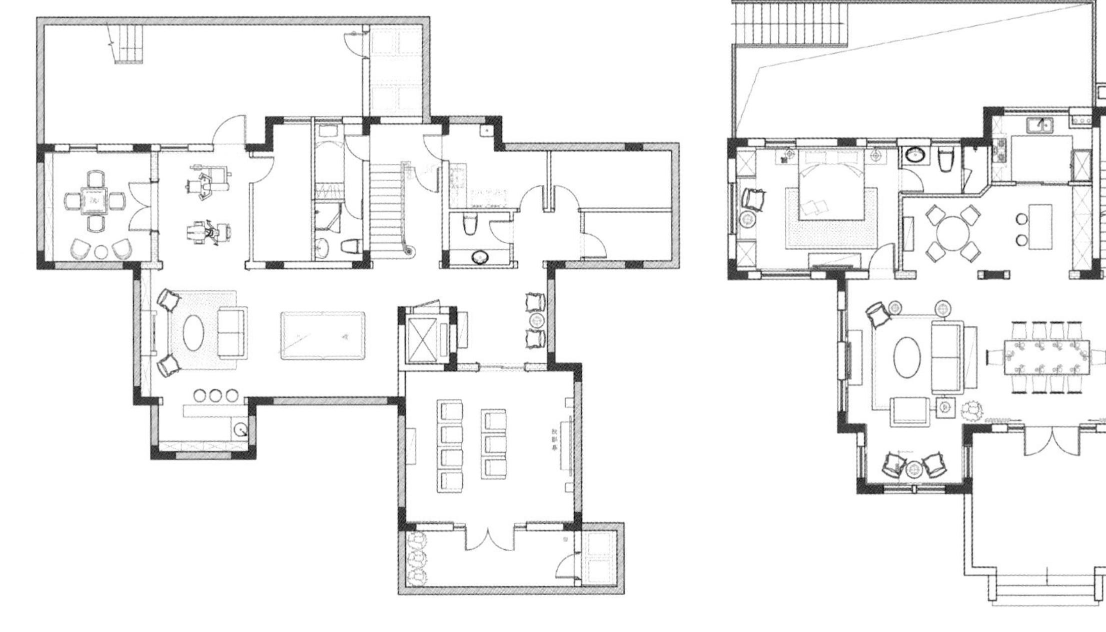

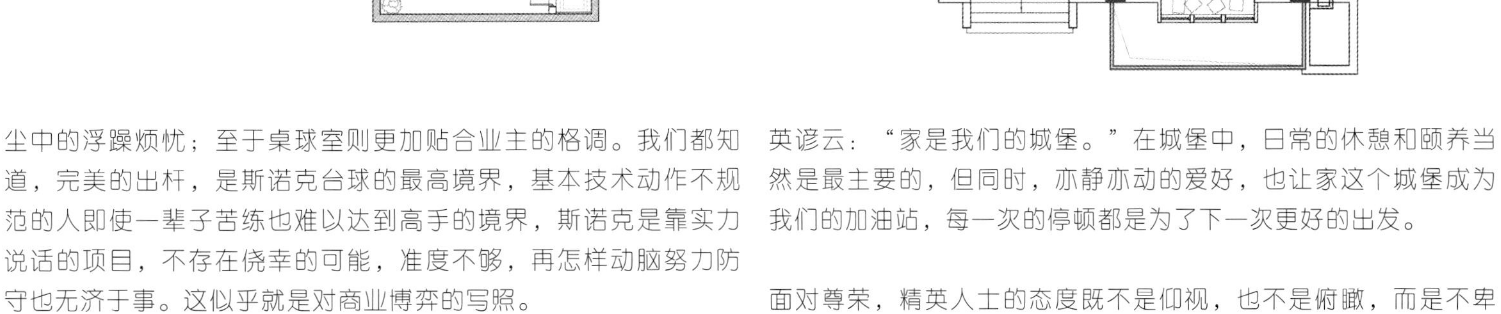

尘中的浮躁烦忧；至于桌球室则更加贴合业主的格调。我们都知道，完美的出杆，是斯诺克台球的最高境界，基本技术动作不规范的人即使一辈子苦练也难以达到高手的境界，斯诺克是靠实力说话的项目，不存在侥幸的可能，准度不够，再怎样动脑努力防守也无济于事。这似乎就是对商业博弈的写照。

英谚云：“家是我们的城堡。”在城堡中，日常的休憩和颐养当然是最主要的，但同时，亦静亦动的爱好，也让家这个城堡成为我们的加油站，每一次的停顿都是为了下一次更好的出发。

面对尊荣，精英人士的态度既不是仰视，也不是俯瞰，而是不卑不亢的平视，本案的设计同样如此；始终关注着客户内心深层需求的贴心服务。

The neoclassical style aims at using modern techniques and materials to restore classical spirit. The case has classical and modern dual aesthetic effect. The perfect combination of ancient and modern gives people spiritual comfort. The case uses new classic style to interpret the heritage of classical humanistic spirit, undoubtedly creating a classic continuity and updates. Everywhere flashes cultural heritage and charm.

The case is a collision and meet of classical and fashionable style. Ionic style Roman column, pile fireplace reflecting the spirit of Western heaven, five silver candelabra representing light and love, large and textured beige marble, and metal textured side tables and furniture, decorative abstract painting, Versace brand elements, all seem unconnected, however, they are perfectly blended together here, highlighting the owners' global vision as an international elite.

Stepping into the living room, the roof's decoration and leather sofas, gray carpet reveals the freedom of American-style. The restaurant's six tables are the standard of western dishes, showing deep family atmosphere. The wall mirrors combination achieves an extension of the sense of space. Designers deliberately separates the different bedrooms using blue or brown segment, forming a different color, different shades of warm feeling; the main tone of the kitchen and bathroom is modern style, some new kitchen equipment is built into the overall kitchen facilities, which is low-key and functional. Sanitary equipment uses the concept of Roman bath, flashing classical romance.

It is particularly worth mentioning the study, wine tasting section and billiard room. The study is concise, the owner as the business leader can deal with the business; In the tasting area, no matter drinking alone or with some friends, one can relax in a little drunk, washing away the impetuous worries. As for the billiard room, it coordinates the owners' style. As we all know, a perfect rod is the highest state of snooker. The one whose action is irregular is very hard to be a master even though through a long time training. Snooker is a program relying on strength. There is no way to win without degree of accuracy. This is the reflection of business game.

An English proverb goes: home is our castle. In the castle, daily rest is the most important thing, but at the same time, the static or movable interest also makes the home of the castle our gas station, every pause is better for the next start.

Facing with honor, attitude of the elites is neither up-look nor down-look, but a plain straight-look. Our case's design is the same. The case is always focusing on the inner service of the deep need of guests.

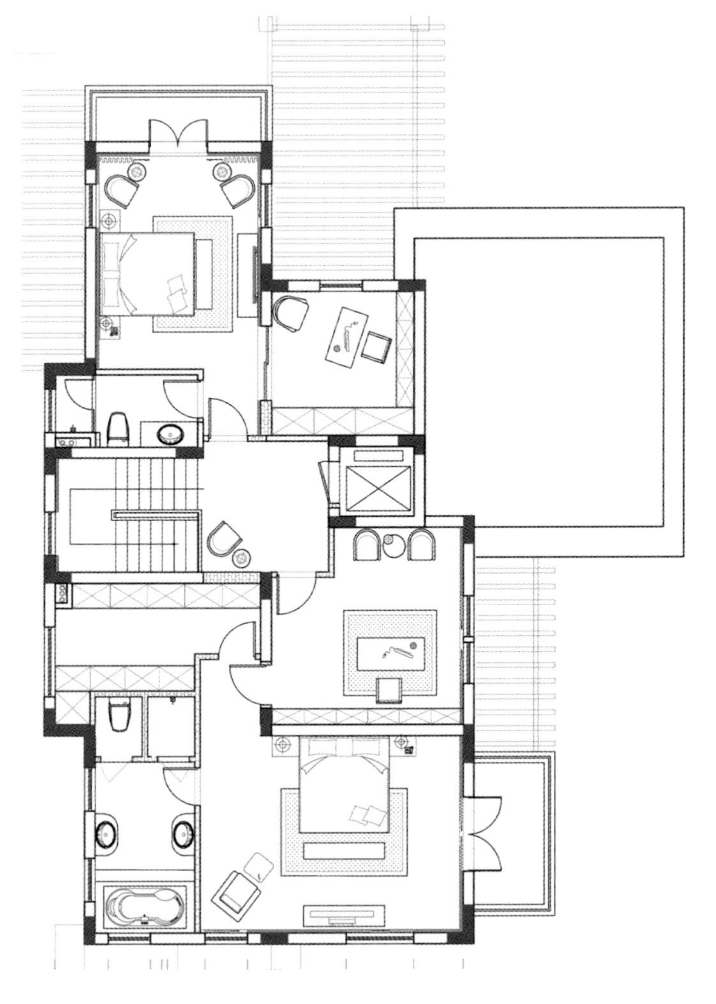

卓弘高尔夫雅苑样板房

ZHUOHONG GOLF YAYUAN SHOW FLAT

设计单位：深圳市尚邦装饰
设计工程有限公司
项目面积：200 ㎡

Design Company: Shenzhen Shangbang Decoration Design
Engineering Co., Ltd.
Project Area: 200 m^2

本案以色彩来诠释时尚现代摩登风格，知性稳重的棕色色调、镜面金属材质的手边、中式元素，勾勒了华丽的视觉效果，让人感受到中西合璧的现代艺术。简洁的线条与规整的家具相互衬托，将古典主义风情融于现代生活，弥漫着强烈的理性色彩和人文关怀。

客厅：中性的色调搭配上鲜艳的配饰和花艺，整个空间简洁明了，干净的线条和丰富的几何图形彰显出智慧、高雅和深沉。以色彩变幻来区分功能空间，巧妙而富有韵味。

书房：木质多功能书柜组合在书房里尽情展现，以国际流行的构图手法将中国传统元素融入室内设计，融现代的律动与中式的沉稳为一体，时尚感十足。

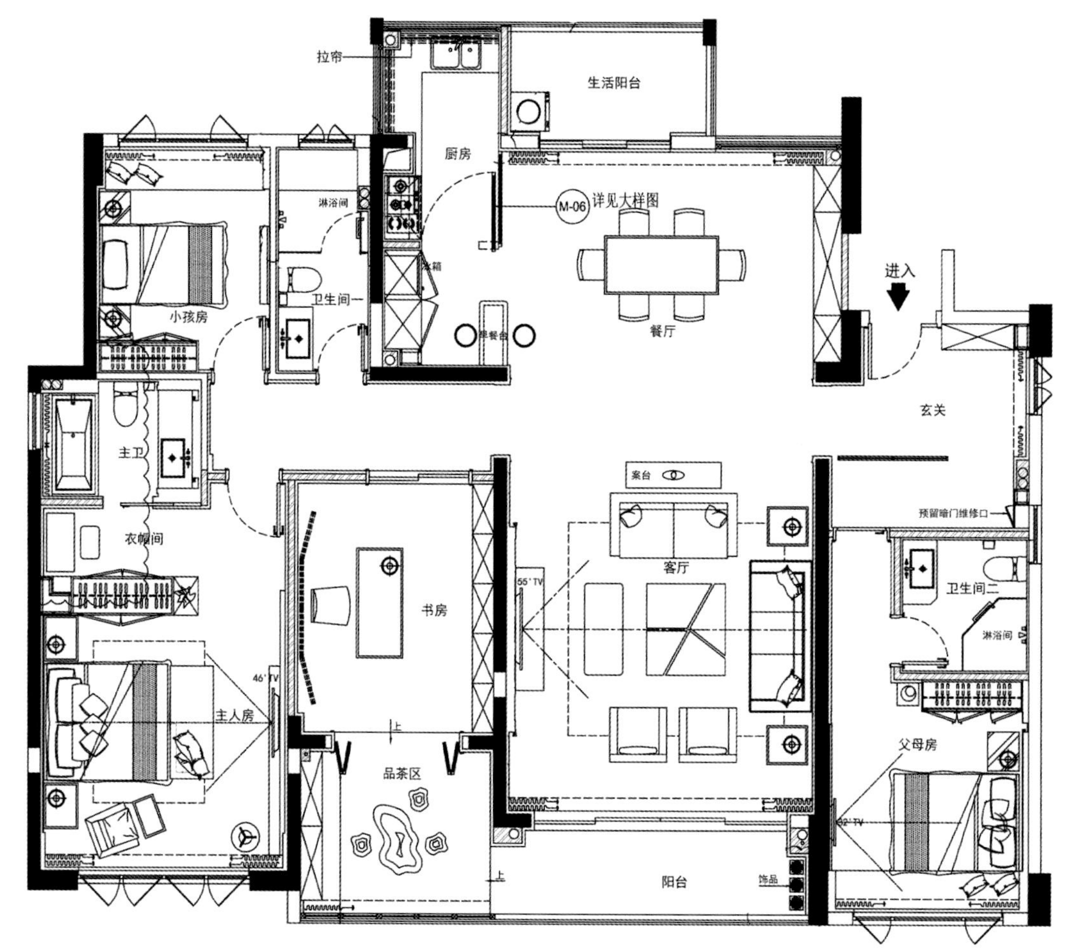

ROCKY
CINDERELLA

CINDERELLA
Dentro di te

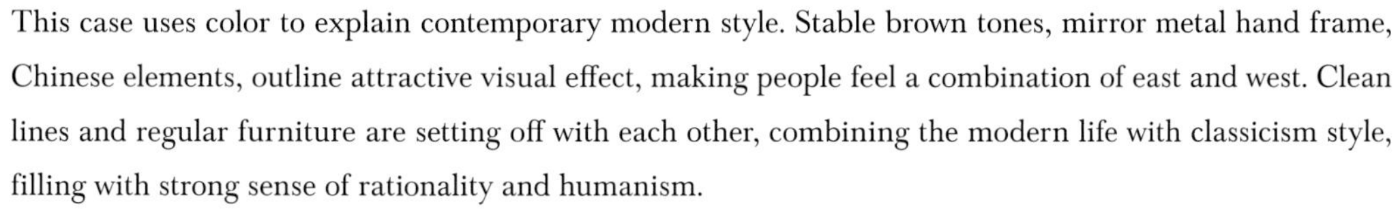

This case uses color to explain contemporary modern style. Stable brown tones, mirror metal hand frame, Chinese elements, outline attractive visual effect, making people feel a combination of east and west. Clean lines and regular furniture are setting off with each other, combining the modern life with classicism style, filling with strong sense of rationality and humanism.

Living room: Neutral color and bright-color accessories make the space simple and clean. Simple lines and rich geometry shows wisdom, elegance and deep. The change color can distinguish the function of the rooms, which seems clever and full of charm.

Study: Wood multifunctional bookcase contains traditional Chinese elements, the international fashion approach combines the elements with the interior design, combining modern rhythms and Chinese fashion sound.

复地朗香别墅

FU DI LANG XIANG VILLA

设计单位：振勇设计事务所
设 计 师：冯振勇
项目面积：630 m²
主要材料：黑胡桃原木、整体木作、圣罗兰大理石、红线米黄大理石、德国美慕厨柜、维宝洁具、汉斯格雅龙头等

Design Company: Zhenyong Design Firm
Designer: Feng Zhenyong
Project Area: 630 m²
Major Materials: Mansion black walnut logs, Wood as a whole, Saint Laurent marble, Red beige marble, Germany Meimu kitchen cabinets, Weibao sanitary ware, Hansgrohe faucet metal, etc.

俗话说得好："物以类聚，人以群分。"我发现不光是本人喜欢木质的厚重感，不少业主也对木材质情有独钟。我和本案业主是2011年接触的，第一套是520m^{2}的别墅，整体采用的是白橡木。本案是我们帮助他打造的第二套独栋别墅了，从设计到施工，再到后期软装搭配，全部是委托我们完成的。在交流过程中，我们有很多的共同点，比如对原木材质的偏爱、对生活实用性的重视程度、对每一位家庭成员的尊重和谦让、对每个人生活中细微差异的包容等，都有着相似的观点。我一直认为，一个好的家装作品不需要去迎合太多人的看法，真的要迎合的是居住者内心的向往，我们要做的其实首要任务就是多沟通交流，挖掘出每位居住者的真正需求，解决空间存在的一些问题，充分发挥出每寸空间的作用。至于风格是否纯粹、造型手法是否独特等，这些问题对于家装设计都不能过于执着，我们拥有一套完整的作品固然重要，但是居住者需要更多的从生活实用性的角度考虑问题。有时候美观与实用出现分歧时，我们必须首选实用性。

The saying goes well that "Birds of a feather flock together". I found that not only I like the weight of wood, a lot of owners also have a deep love towards wood. I know about the case's owner in 2011 and the first set villa locates 520 m^2, the whole material is white oak. This case is the second villa that we help him to build. Design, construction and later soft decorations are all designed by us. In the process, we have a lot in common, such as preference of raw wood, the practical importance of life, the respect to each member of the family, courtesy of subtle differences in everyone's life, etc.. I've been thinking that a good home decoration work does not need to cater too many people. It really needs to meet the occupants' inside. What we have to do first is to have more communication, and dig out the real needs of each residence, solving problems of every inch of space and making use of every inch of space. House design should not stick on the style or approach. It is certain important to create a complete work, while the residents emphasis on the practical use of it. Sometimes when beauty and practice has conflicts, we must choose practice.

东湖上郡

EAST LAKE COUNTY

设计单位：苏州大斌空间设计事务所
设 计 师：杨飞
项目地点：安徽芜湖
项目面积：230 m^2
主要材料：仿古砖、爵士白大理石、银镜、壁纸、木线条等
摄 影 师：AK 空间摄影

Design Company: Suzhou Big Bin Space Design Firm
Designer: Yang Fei
Project Location: Wuhu, Anhui
Project Area: 230 m^2
Major Materials: Antique brick, Jazz white marble, Silver mirror, Wall paper, Wood lines
Photographer: AK Space Design

本案为一套复式住宅，地处安徽芜湖。从原始户型即可看出整个房子的动区和静区划分很合理。进门左手边为厨房餐厅，正中间为客厅，右边为卧室。玄关的左手边原本有强、弱电箱，因此做了一个“回”字形的镜子门板，并安装了穿衣镜，这样可将强、弱电箱都隐藏在门后方。玄关是进入房子的第一视角，几何图形的亮色壁纸很好地呼应了入户楼道窗口的铁艺雕花，同时也提亮了进入玄关的视觉冲击感。

入户玄关位置：地面选择了深咖啡、海蓝色、玛莎拉红、白色的混拼，一来层次分明，二来耐脏。设计师在玄关处增加了一处视觉缓冲的“屏风”线条拼嵌镜面，不仅可以借景，还可以在出门之前整容。

挑高的客厅是整套房子的亮点，同时也是设计的一大雕琢点。在设计过程中虚实结合，考虑了视觉空间的对称性，电视背景的左边拱形门洞和右边的拱形镜面是对称的，沿着客厅的中心线将电视背景横向折叠，你会发现，原来上下也是对称的，中间采用爵士白大理石，起到了一个大立面很好的过渡作用。皮质沙发配搭布艺装饰，从顶到地的波纹幔头窗帘将整个空间融为一体。

从电视背景的左手边门洞进入便是餐厅区域，原本狭小的厨房向外扩大，使餐厅减去餐边柜之后成为了一块方形区域，使之具有宽敞感，美式深色圆桌增加了就餐区域的空间感。厨房深浅点缀的仿古砖与餐边柜区域的墙面呼应，使整个餐厅、厨房形成了统一的整体。

卧室之间的共享小空间，选择了和电视背景材质一致的爵士白大理石，象牙白色的门套，加高的门套线，浅烟灰色纯色壁纸，立体构成的麋鹿造型墙饰，就像一幅山水画。

预留给未来第二个孩子的房间，尽管不知道是男宝还是女宝，选择中性色也不会显得突兀。

楼梯是一个衔接楼上楼下的阶梯。图书馆的楼梯可以让你坐下来沉迷书海，校园里的楼梯可以坐下来谈天说地。本案楼梯用爵士白的大理石配上波纹的壁纸、经典的色调，使两个空间完全融合为一体。在这个信息化的时代，人与人之间面对面的交流越来越少，所以在靠近栏杆的地方放置两张休闲椅，坐在楼上的休闲区可以和楼下的亲人朋友很好地互动交流。

The case is a suite of duplex apartment, it locates at Wuhu, Anhui Province. From the original unit, one can see that the dynamic and quiet areas are divided reasonably. The left side in the door is kitchen and dining room, the middle is living room, the bedroom is on the right side. The left side of the entrance has strong and weak electronic boxes. Therefore the designer makes a hollow shape mirror doors. They install fitting mirrors behind the door. Therefore the strong and weak electronic boxes are hidden behind the door. Entrance is the first visual perspective of the house, the geometry light color wallpaper well echoes the carved iron flower at the entrance of corridor window, meanwhile enhancing the visual sense of the porch.

Entrance porch location: The ground uses dark coffee, deep aquamarine, Martha red,

white blending. The match is layered and stain-resistant. Designer adds a visual mirror at the entrance of "screen", before the mirror one can not only see the view but also clean the face before going shopping.

The living room's height is the highlight point of the whole house, and also a big carved point of the design. The design process combines the fiction and reality design, taking into account the symmetry of the visual space. The arched doorway on the left of the TV background and the right arched mirror of the TV background are symmetric. Along the centerline of the living room to horizontally folds the TV background, you'll find the up part and the down part are symmetrical either. The jazz white

marble façade has a good transition effect. The leather sofa matches fabric decoration, the corrugated mantle curtain from top to the end combines the spaces together.

Entering into the house from the left of door of the TV background, the small kitchen expands outward. The restaurant minus the side cabinet becomes a square area, the space is full of spacious feeling. American dark round table increases the sense of space in the dining area. Kitchen's antique brick echos with the side cabinet wall. The kitchen and the dining room becomes an unified whole.

The share small space between the bedrooms chooses the same Jazz white marble of TV background materials. The white ivory marble, high door line, pale smoky gray color wallpaper, and stereoscopic Elk shapes wall decorations are like a complete landscape painting.

Though do not know the gender of the second child, a neutral color will not be improper in he/she's room.

Ladder stairs is a connection that connects upstairs and downstairs. One can sit in the book sea on the library's stair, one can sit and chat on the school's stair. The stairs in the case use jazz white marble and corrugated wallpaper. The classic colors connect the two spaces together. In this information age, face-to-face communication between people is less and less, so designer places two lounge chairs close to the railing, sitting at the upstairs lounge one can talk with the family and friends downstairs.

怡景苑

YI JING YUAN

设计单位：深圳鸿艺源建筑
室内设计有限公司
设计师：郑鸿
项目地点：海南海口
项目面积：190 m^2

Design Company: Shenzhen Hongyiyuan Architecture
Interior Design Co., Ltd.
Designer: Zheng Hong
Project Location: Haikou, Hainan
Project Area: 190 m^2

本初　试想，生命之初，万物归于道。继而道生一，一生二，二生三，三生万物。大千世界，芸芸众生，得此而来。自然之道，乃道之根本，故自然乃万物本初。

渐远　邵华一生，岁月留痕行色匆匆，繁华都市，你我都在迁流不居中自得一隅。回头看来时的路，离那份稚子之心，已渐行渐远。自然早已成为一种奢侈谜尚，知遇难求。

归来　回归，是一种态度。羁鸟恋旧林，池鱼思故渊。人，不忘初心，方得始终。繁花满树，天心月圆，水随天去，定有攸归。

本案以简欧风格为主打，融入美式风情和东南亚民族特色，抛开金碧辉煌的外在形式，褪去华丽烦冗的形式语言，力求用简单的表现手法营造最大的舒适空间，将重点投入到每样物品本身，使其散发出自身独特的魅力。简洁不只是单纯的朴素和节俭，而是能超越豪华奢侈的简约。

“素”是一种拙朴的形态。“素雅”则是一种心境，简洁而不累赘。层峦万象，只虏一景；弱水三千，只取一瓢。

留下一片玉壶冰心，野鹤闲云，质朴如初。没有自我的强调，也就没有了表达的负担，降低高调的姿态，收获返璞归真的从容。有无相生，换得一种随心，一种任性，一种自在悠游。

岁月辗转，几经沉淀。

如今，看山还是山，看水还是水。

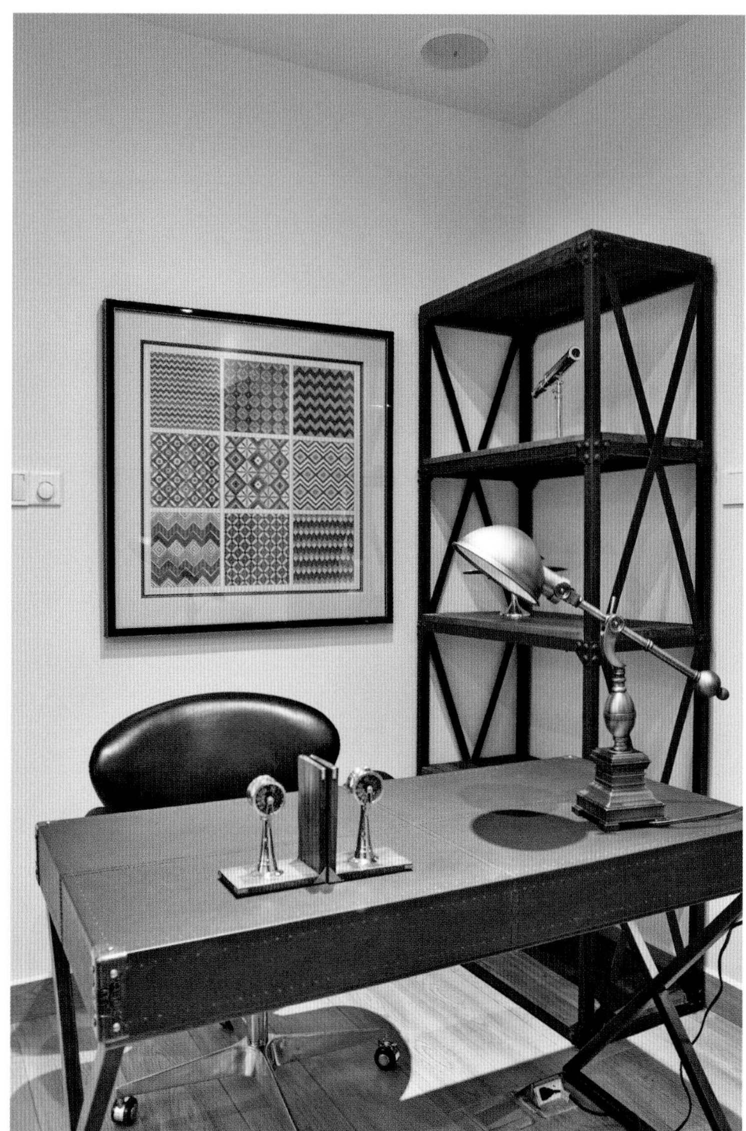

Beginning At the beginning of life, everything belongs to Tao. Tao creates all things in the world. Nature Tao is the origin of Tao, therefore, nature is the beginning of all things.

Far away The whole lifeis hurry and bustling. City people randomly live in the prosperous place. Looking back at the former road, we are far away from the heart of the beginning. Nature has become a precious luxury and mystery fashion.

Return Return is an attitude. Bird loves old forest, fish loves old pool. The beginning heart makes things satisfactory. Flowers blossom on the tree, sky is bright, moon is round, water follows downwards, and will return one day.

The case uses simple European style as the main style, combining with American style, and Southeast Asian national characteristics.Throwing away the golden form and gorgeous burdensome language, designer strives to create maximum comfort space with a simple expression way. They focus on the materials, makethem exudes their unique charm. Simple is not merely frugal, but beyond the luxurysimplicity.

NEW YORK

"Su" is anaustere form. "Su Ya" is a kind of inner status-simple but not cumbersome. The beautiful mountains are all there, but I just picture one of them. The ocean is wide, but I just take one cup of water.

Finally, I leave my heart in the ice pot. The wide crane flies in the cloud, the sense is plain and pure. The lose of oneself means lose of expression. People reduce their high attitude, and will gain pure easy feel. Life will turn into a kind of random and freedom.

Time flies, time precipitates.

Now, mountain is still mountain, water is still water.

招商雍景湾

ZHAO SHANG YONG JING BAY

设计单位：苏州大斌空间设计事务所
设 计 师：曹亮
参与设计：大斌空间设计团队
项目面积：190 m²
主要材料：涂料、壁纸、木饰面、抛光砖、马赛克
摄 影 师：AK 空间摄影

Design Company: Suzhou Big Bin Space Design Firm
Designer: Cao Liang
Participatory Designer: Big Bin Space Design Group
Project Area: 190 m²
Major Materials: Painting, Wallpaper, Wood finishes, Polished tiles, Mosaic
Photographer: AK Space Photography

Paramount
100

这是一个190 m^2的大平层，首先进入门厅的是较弱的光线，在设计上没有去提高亮度，反而用了一个相对较深的色调去减弱光线，我们希望当业主穿过这个空间进入客厅或者房间的时候，光线突然放大，通过这种明暗的反差让空间能够变得生动一些，或者说更加有趣，同时能够形成区域上的明显分界。

这个空间是一个四室的户型，根据其家庭人口，我们调整成了3+1的空间，所以可以看到在主卧大区域有一个套间的构成，独立的卫生间、衣帽间和书房，当独立的卧室改造后，我们通过两扇到顶的移门，又把空间变成了四房的格局，通过切换，使每个空间实现最大的利用率。

客餐厅区域大面积的落地窗户，光线本身就是最好的装饰物，所以硬装上不需要去刻意做造型，只是通过颜色、材质、灯光来体现要表达的风格。

在色彩的运用上进行了巧妙搭配，大空间用了两种不同的灰色去做背景色，灰色本身属于后退色，能够放大空间，同时在软装饰品上用同样的颜色去呼应协调背景色，简单的搭配有效地增加了空间的层次感和立体感。

This is a 190 m^2 big flat layer, the first come into the hall is weak light. Designers do not enhance brightness on the design, instead, they apply a relative deep tone to weaken the light. We hope that when owners walk through this space and get into the living room or rooms, the light suddenly enlarges, through the contrast of light and dark, designers make the space vivid and interesting. Thus regional boundaries has been formed.

This space contains four rooms, according to family population, we adjust the space into 3+1 type. We can see there is a suite at the main area of the bedroom, including a bathroom, cloakroom and a study. After the transformation of bedroom, we use two moved door and change the space into a four rooms pattern, through switch, we make each space achieved maximum utilization.

As for the large area of windows, light itself is the best ornament, therefore we don't need to deliberately style the house. We just use color, texture, and light to reflect the expression style.

We use clever collocation on the use of color. The large space has two different gray background, gray itself belongs to receding color. It can expand the space. And the designer uses the same color on the soft decoration to echo the background colors, the simple collocation effectively increases the level of space and three-dimensional sense.

绿湖豪城

GREEN LAKE BRILLIANT CITY

设计单位：方块空间工作室
设计师：蔡进盛
项目地点：江西南昌
项目面积：510 m^2
摄影师：邓金泉

Design Company: Square Space Interior Firm
Designer: Cai Jinsheng
Project Location: Nanchang, Jiangxi
Project Area: 510 m^2
Photographer: Deng Jinquan

Birds

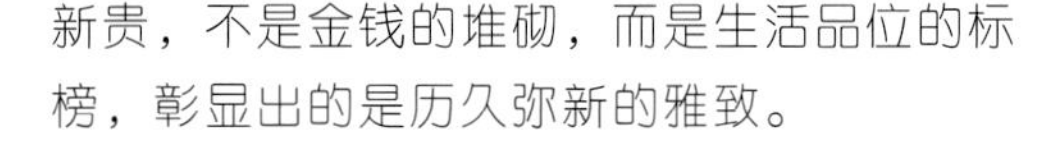

新贵，不是金钱的堆砌，而是生活品位的标榜，彰显出的是历久弥新的雅致。

坐落于红谷滩的绿湖豪城别墅，整体地貌丰富，生态良好、交通便捷。区内自然环境优越，绿树成荫，坐拥两湖，活水、网络纵横交错。没有喧嚣，没有污染，没有厚重的工业感，是城市里尚未开发的净土，又仿佛是沉睡的瑰宝。这个风格鲜明的别墅区为住户提供了临湖而居的怡人的生活环境，是高端业主选择宅邸的最佳地理区域。

面对生态悠然的居住环境，室内的装修就显得更为重要了。新古典主义的风格总是备受人们的喜爱，本案将古典风韵与气质完美地融于一体，传统纹样与现代材质的结合、古

典造型与时尚形式的重新演绎，加之后期配饰的点缀，使本案又多了一些极具贵族气质的ARTDECO风格。全新的生活理念让业主始终能够从容地享受物质生活，同时更能享受精神层面的愉悦。

餐厅中多层次的方形吊顶，配以浅色酒柜，与镶嵌方形边框的地面石材相呼应，自然地围合出餐厅的空间。

本项目为3层单体建筑，一层为大堂、接待处；负一层为酒吧、VIP洽谈区、雪茄房、投影区、KTV；二层为宴会厅、办公区等。把英伦和法式这种古典复兴的特点引入室内设计中，用现代的处理手法打造成既重实用又不失时尚典雅的设计典范。

The new noble is not a pile of money but a flag of life taste, showing timeless elegance.

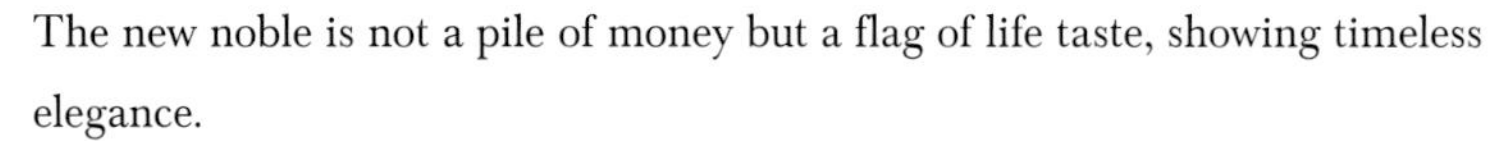

The Green Lake city villa on Red Valley Beach has rich landforms, fine ecology and easy transportation. The natural environment is beautiful including lush trees, two lakes, crossing water. There is no noise, no pollution, no heavy industry. It is a virgin land in the city. It is also like a treasure. The villa district provides beautiful life beside the lake. It is the best place for high-end owner.

Facing the ecological healthy living environment, inner decoration is more important. New classic style is always loved by people. The case combines the classic atmosphere with the perfect temperament. The combination of tradition and modern material, replay of classic and fashion style as well as the ornaments adds more noble ARTDECO style. New life idea makes the owner can always enjoy life physically and mentally. The square ceiling in the dining room are matched with light cabinets, echoing with the stone with square frame, naturally creating dining room's space.

The program is a three-storey building. The first floor contains lobby room and reception space. The underground floor contains bar, VIP negotiation area, cigar room, projection area, KTV. The second floor contains banquet hall, office etc. Designer brings English and France classical style into the interior design, using modern way to create a practical and fashion design style.

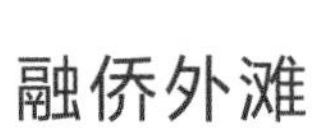

融侨外滩

RONG QIAO BUND

设计单位：品川设计
设 计 师：林新闻
项目面积：310 m^2

Design Company: PURE CHARM
Designer: Lin Xinwen
Project Area: 310 m^2

提起传统风格，映入脑海中的第一个印象或是沉重的中式，或是线条繁复的古典欧式，而在本案中却完全找不到一件类似的家具。设计师摒弃了华丽复杂的装饰，以当代简约的审美在线条上做减法，使得极简成为最美的风格。巧妙运用淡雅的色彩和轻盈的线条营造出富有品位和身份的空间，最终让家回归到最本质的纯粹、简单。

根据生活习惯规划客厅和餐厅，空间不做完全封闭的间隔，而是完全敞开，更显通透、开阔、完整，有效地提升了居室空间的利用率。地面的处理很简单，棕色的木纹打底，光润整洁。墙体和天花板全部运用象牙白和纯白，选取的家具也是脱胎于传统风格的改良样式，没有过多的色彩，线条平直简洁，彻底摆脱了迂腐和匠气。简约的空间延续整体风格，对称式的格局布置让家增添了一份整洁而又严谨的仪式感，适当地利用精致的物件点缀空间，无形中也让家的气质有了更高层次的升华。

如果一个家的公共空间还承担着展示给他人的社交功能，那么卧室则是全然私密性的。与公共区域的四地落白相比，卧室的色彩则多了一份内敛气质的法式融情，从墙面色彩到软装搭配，或温馨淡雅或质朴无华，简简单单的设计反而带出“少即是多”的感官感受，给人留下了大家风范的美好印象。一个空间如果全都是纯白粉刷，难免显得过于二维平面化，本案的设计师在简单的手法之中，利用色彩和简约的线条将精致深化到一幅画、一个摆件，抑或是家具的曲线上，以此加深空间的层次感。细节之处让这个家增添了时尚、高雅的品位。

As for traditional style, the first impression comes to mind is heavy Chinese style, or classical European style with complex lines. However, in this case, one can not find a similar piece of furniture. Designer rejects magnificent complex decorations, using contemporary minimal aesthetic to do subtraction, makes minimalism the most beautiful style. Designer cleverly uses elegant colors and graceful lines to create a rich taste and identity space, eventually makes the family returning to the most essential, pure and simple status.

According to habits and customs, designers create the living room and dining room. The space does not a total closed interval, but totally open, which seems more transparent, open, complete, thus effectively improve the utilization of living room space. Ground design is very simple, the brown wood color is used as bottom color, makes the house smooth and clean. All

the walls and ceilings are ivory and white. The style of the selected furniture is also based on a traditional style. The design has no more color, the lines are straight and simple, totally get rid of the pedantic thinking and triteness. Simple space continues the overall style, symmetrical pattern adds a neat and precise sense of ceremony. Designer appropriately uses exquisite objects to decorate the space, enhancing the home's layer.

If the public space of a home shoulders the social function of display, then the bedroom is totally private. Comparing with the public areas' white floor, the bedroom's color has a piece of introverted French temperament. From the wall color to the soft decoration, the temperature is warm and elegant. Simple design has "less is more" sensation, give people a good impression of the style. If a space is all white, it inevitably seems too two-dimensional. Designers of the captioned case use simple approach, colors and simple lines to deepen the delicate sense into a drawing and an ornament or a curl of furniture. This way deepens the layer of the space. The detail adds fashion and elegance to the home.

上林西苑叠拼别墅

SHANGLIN WEST YUAN SUPERIMPOSED VILLA

设计单位：上海无相室内设计工程有限公司
空间设计师：王兵、王建
软装设计师：李倩
项目面积：210 m²
主要材料：仿古砖、榆木实木、乳胶漆、特殊漆、石材马赛克
摄 影 师：三像摄 张静

Design Company: Shanghai Wuxiang Interior Design Engineering Co., Ltd.
Space Designers: Wang Bing, Wang Jian
Soft Designers: Li Qian
Project Area: 210 m^2
Major Materials: Antique brick, Elm wood, Latex paint, Special paint, Stone mosaic
Photographer: Three photographer Zhang Jing

也许是看透了镁光灯辉煌之后的落寞，人生历练和对生活真谛的谙熟，浮华之后尽显自然的可贵。某种记忆，在内心深处总是不可磨灭的存在。无论外界如何的物欲横流，如何的珠光闪烁，总有一个自然、古朴的世界是属于你的。让我们把它放大，加以表达吧，如榆木实木上的年轮，将自然的世界充分展现出来！此案注重利用粗犷墙面表面肌理和质感来营造一种自然质朴温馨的居家气氛，采用自然的天然材质，辅以自然的色泽，给空间以清丽恬静的感觉。纯铜的吊灯配以金边色的手稿画相映成趣。在家具方面注重原木色和真皮质地材料的搭配，体现了主人对自然世界的向往。

After seeing through the loneliness behind the brilliant spotlight, after getting many life experience and true life meaning, one will cherish the value of nature. Some memories deep in the heart always been indelible. No matter how materialistic the world is, there is always a natural, ancient world belongs to you. Let us expand it and express it! Let us use Elm wood rings to express the world! The case focuses on the rough wall surface texture to create a natural and plain warm home atmosphere. The house uses natural material, natural color, to give the space a sense of beauty and quiet. Copper chandeliers and golden frame manuscript painting are echoing with each other. The furniture focuses on natural color leather texture materials, reflecting the owner's yearn for the natural world.

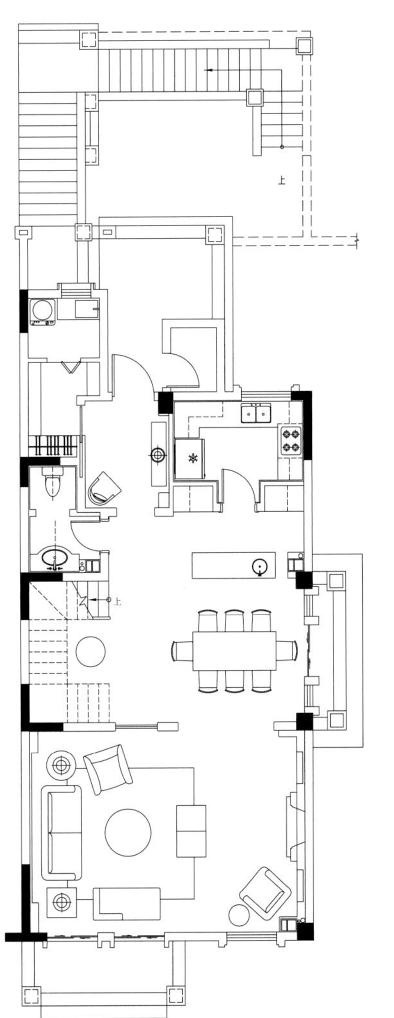
上
上

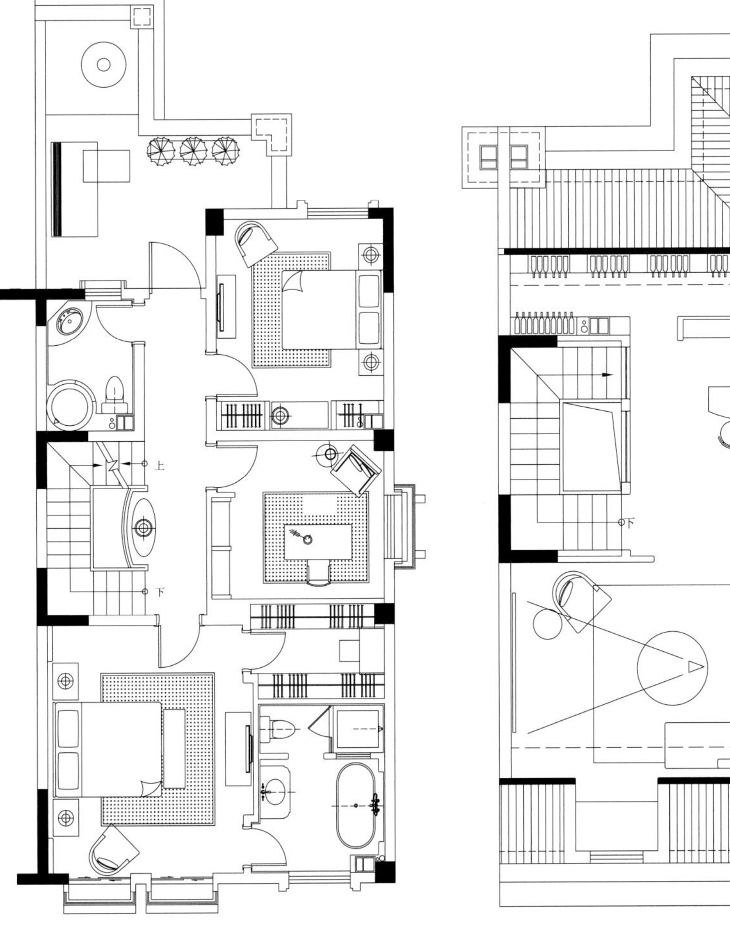
上
下

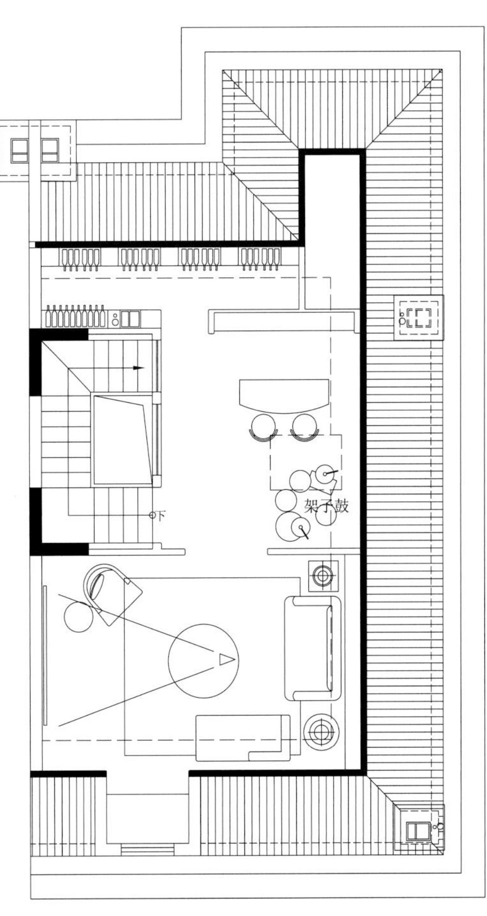
下
架子鼓

PARAPLUIE-REVEL
LYON
(FRANCE)
London Evening News
London
London
London
3
4
2

本书在编写过程中，得到各位参编老师的倾力协助，特表示感谢，
以下为参编人员名单（排名不分先后）：

唐震江 郭　颖 吕荣娇 欧阳云 张　淼 王丽娜 王　寅 夏辉磷 华　华 贾　蕊 廖四清
葛晓迎 高　巍 张莹莹 张　明 张　浩 梁敏健 邓　鑫 刘升山 刘　斌 许友荣 郇春园
史樊英 史樊兵 吕　源 吕荣娇 吕冬英 张海龙 段栋梁 孙朗朗 张　艳 李美荣 陈靖远
宋献华 吴源华 朱臣高 鲍　敏 刘勤龙 付均云 胡荣平

图书在版编目(CIP)数据

诠释豪宅 / ID Book图书工作室编. --武汉：华中
科技大学出版社，2016.1
ISBN 978-7-5680-1228-7

Ⅰ. ①诠… Ⅱ. ①I… Ⅲ. ①住宅－建筑设计－图集
Ⅳ. ①TU241-64

中国版本图书馆CIP数据核字(2015)第222317号

诠释豪宅　　ID Book图书工作室 编

出版发行：华中科技大学出版社（中国·武汉）
地　　址：武汉市武昌珞喻路1037号（邮编：430074）
出 版 人：阮海洪

责任编辑：胡　雪　　责任监印：秦　英
责任校对：曾　晟　　装帧设计：张　艳

印　　刷：深圳当纳利印刷有限公司
开　　本：965 mm×1270 mm　1/16
印　　张：20
字　　数：288千字
版　　次：2016年1月第1版第1次印刷
定　　价：338.00元(USD 53.27)

投稿热线：(010)64155588-8000
本书若有印装质量问题，请向出版社营销中心调换
全国免费服务热线：400-6679-118　竭诚为您服务